International Political Economy Series

Series Editor
Timothy M. Shaw
Visiting Professor
University of Massachusetts Boston
Boston, MA, USA

Emeritus Professor, University of London, London, UK

The global political economy is in flux as a series of cumulative crises impacts its organization and governance. The IPE series has tracked its development in both analysis and structure over the last three decades. It has always had a concentration on the global South. Now the South increasingly challenges the North as the centre of development, also reflected in a growing number of submissions and publications on indebted Eurozone economies in Southern Europe. An indispensable resource for scholars and researchers, the series examines a variety of capitalisms and connections by focusing on emerging economies, companies and sectors, debates and policies. It informs diverse policy communities as the established trans-Atlantic North declines and 'the rest', especially the BRICS, rise.

More information about this series at
http://www.palgrave.com/gp/series/13996

Clare O'Grady Walshe

Globalisation and Seed Sovereignty in Sub-Saharan Africa

Clare O'Grady Walshe
School of Law and Government
Dublin City University
Dublin, Ireland

International Political Economy Series
ISBN 978-3-030-12869-2   ISBN 978-3-030-12870-8   (eBook)
https://doi.org/10.1007/978-3-030-12870-8

Cover image © Rob Friedman/iStockphoto.com

This Palgrave Macmillan imprint is published by the registered company Springer Nature Switzerland AG.
The registered company address is: Gewerbestrasse 11, 6330 Cham, Switzerland

*To my adorable children, Caoimhe and John; to the work of
seed savers everywhere; and to the memory of those who died and continue to
die in needless famines in a world of abundance, beauty and diversity.*

PREFACE

This book was inspired by my background in environmental, justice and human rights work. From early days as a toxics campaigner for Greenpeace International, to country director of Greenpeace in Ireland and trustee of Greenpeace International Council in the 1990s to my work with Irish Seed Savers since 2008, I have a passionate interest in these interrelated issues, which affect people and planet and determine the future of both.

Greenpeace involvement was instructive on a number of counts. It brought me to some of the most beautiful places on the planet and to some of those that have been the most devastated and destroyed by pollution. I was involved in various campaigns associated with the negative effects of intensified farming and chemicalisation of agriculture as the dangers of the corporate takeover of biological systems were starting to emerge. Of particular concern at that time was the introduction of Genetically Modified Organisms (GMOs), following the Agreement on Agriculture and big chemical/pharma interest moving into seed and genetic acquisition of plant species. There was already alarm that various globalising institutions were being formed alongside increasingly powerful transnational corporations (TNCs) that were significantly changing the rules, accountability and ownership of the agricultural landscape with potentially negative consequences for people and environment. This period also saw increasing public mobilisation and deepening awareness of the dangers of biodiversity and species loss exacerbating the effects of climate change. It was finding more traction everywhere. By 2005 and 2010 leading scientists across the globe reported in the UN millennium ecosystems assessment reports that two-thirds of the earth's ecosystems were in danger of collapsing, thus undermining the possibility of achieving the solemnly declared UN Millennium Development Goals, the precursors to the Sustainable Development Goals (SDGs). I had the opportunity to work on these issues at a national level through my work on the Irish Aid Advisory Committee, the Heritage Council and the National Task Force on Green Enterprise. In the latter, I successfully pushed for the inclusion of biodiversity in the text of the final document. This was a significant inclusion as it signalled a willingness at the state level to accept that biodiversity and the ecological system are the bedrock of a healthy economy and society into the future. My insistence on its importance and inclusion came from having become a grower and a seedsaver myself.

In the early 2000s, having started a family in a rural village in Ireland, we dug up the back garden and began growing heritage varieties of Irish seeds from the Irish Seed Savers Association (ISSA). ISSA is a small charity dedicated to the conservation and restoration of heirloom varieties of predominantly, though not exclusively, native seeds. I was enthralled to observe the capacity and abundance of locally adapted seeds, comparatively to other seed varieties. Some of these seeds had been repatriated from as far away as the Vavilov seed bank in St Petersburg in Russia.

The work of Irish Seed Savers opened my eyes to the coevolution of people and seeds throughout human history. I began to see a pattern emerging of new global institutions alongside powerful geopolitical allies determining seed and therefore food futures on every continent, sometimes undeterred, indeed assisted by the precarious fragility of the state. This manifested itself in the war zones of Iraq and later Afghanistan, where seed laws were completely changed during times of military occupation, to venture philanthropists pushing GM and commercial seed agendas on cash-strapped, poverty-compromised countries in the Global South. It appeared that a new global seed order was being shaped through a highly sophisticated means of appropriating the legal tools of international law to usurp local seeds and establish commercial seed systems, which greatly benefit corporate seed interests.

I was invited to join the board of directors at ISSA in 2008 and had the privilege of introducing the President of Ireland, Michael D Higgins, to open the first community seed bank in Ireland at Irish Seed Savers in 2011. This seed bank, adjoining the living library of a 20-acre farm, now constitutes an important state and international collection of non-commercially available fruit and vegetable seed, numbering 800 varieties, mostly Irish native varieties. The value of restoring such a community seed collection cannot be underestimated. Ireland, which suffered horrific famine in the 1840s when there was a dependence on predominantly one variety of potato, 'the lumper', within a monoculture cropping system during colonial times, continues to provide a salutary warning to the world. Replicating such community seed initiatives everywhere now is surely the most wise, hopeful and practical act of recovery, restoration and conservation in the face of climate change, hunger and biodiversity loss.

Dublin, Ireland   Clare O'Grady Walshe

ACKNOWLEDGEMENTS

I would like to thank the team at Palgrave Macmillan for supporting the publication of this book. They have encouraged me every step of the way to transform the core research of my PhD into a comprehensive book that can reach a wider audience. Their commitment and professionalism has been unstinting from start to finish. Special thanks to Sarah Roughley and Oliver Foster who organised everything so smoothly at Palgrave Macmillan and Professor Timothy Shaw for welcoming me and my work into the International Political Economy (IPE) series.

I want to thank those wonderful people who have funded this post-doctoral fellowship and made this book possible. Special thanks to The Frank Jackson Foundation, in particular Pete Brown, who was so enthusiastic about my research in East Africa and was the first to come in behind this project. Renowned Chef Darina Allen and the Slow Food Movement in East Cork supporting my project meant so much to me, as it was in Cork that my own involvement in environmental concerns began. I am deeply grateful to Matt Dunwell of Ragman's Farm in the UK and to Action from Ireland (AfrI) and to other private donors whose support for me has been so encouraging.

My thanks to the School of Law and Government in Dublin City University (DCU) for awarding me a scholarship to pursue a PhD, which is the basis for this book. Thanks especially to Professor Gary Murphy, former Head of School, for his kindness and support throughout the years, and to so many members of staff and student body who have helped and stimulated me during my time at DCU.

My deepest gratitude goes to Professor Robert Elgie, who provided such excellent supervision of my PhD and remains a dedicated mentor for my work. I am so grateful for his unrelenting patience, respect and professionalism. It was a pleasure and an absolute privilege to work with him. He consistently assisted me in bringing the best academic rigour to this work and his commitment and unwavering support means so much to me.

I am especially grateful to Anita Hayes, who literally sowed the seeds for this book through her inspirational work as the founder of the Irish Seed Savers Association and introduced me to this important area of enquiry and to a wonderful team of people there. She and her husband Tommy have encouraged me every step of the way. Thanks also to all the staff over the years at Irish Seed Savers, especially Jo Newton and to Matteo Pettiti for scientific assistance over the years and for introducing me to the inspirational work happening across Europe, particularly with Rete Semi Ruralis in Italy. Thanks also to Dr. Paul Dowding, Fellow Emeritus Botany, Trinity College Dublin Centre for the Environment, and Matthew Jebb, Director of the National Botanic Gardens, Glasnevin, Dublin, for their encouragement along the way.

Special thanks to Dr. Kathy Glavanis Grantham, thank you for your enduring friendship and support over all the years since my earlier days at University College Cork and for your generosity in introducing me to so many wonderful Ethiopian academics, who assisted me during my fieldwork, especially at Mekelle University, and in Addis.

Huge thanks to all the people who helped me in Kenya and Ethiopia. To my wonderful friends, Mary and Gary, I give special thanks. You opened every door for me and shared your lovely home with me in Ethiopia. It hugely enriched my work. Thanks to so many people at the various agencies and non-governmental organisations (NGOs) who generously gave of their time. I am especially grateful to the Gaia Foundation for introducing me to Dr. Melaku Worede, who is not simply a teacher. He is rightly regarded as 'A School'. Together with Regassa Feyissa and Dr. Sue Edwards they introduced me to the rich and diverse agrarian world of Ethiopia. Thank you all for your time and invaluable help. Thank you to Dr. Fetien Abay at Mekelle University for her time and her introduction to her barley research programme in the Highlands. My special thanks to all the wonderful people of the communities who welcomed me into their homes and co-operatives in Oromia and Tigray.

To the wonderful people of various NGOs who helped me along the way, especially Patricia Wall and all the team at Trócaire and to the team at

Self Help Africa in Ethiopia, who provided me with so much help and information, I am indebted to you all. Thanks to Brian Daly who introduced me to his many colleagues at Diageo and for all their assistance.

Thanks to my true friends, Joe Murray of AfrI and Richard Moore of Children in Crossfire for all your support. Your introductions ensured that I was well looked after in Nairobi by Sean Cremin and P.J. McCamphill of St Patricks Mission, and by Gebremichael Ghembera of Children in Crossfire in Ethiopia who became my guide, translator and trusted friend. I owe special thanks to Daniel Maingi of Kenyan Food Rights Alliance (KEFRA) for so much assistance and insights and to all of the Kenyan parliamentarians for your openness and hospitality at the Kenyan Parliament.

Special thanks to the true strategic alliance that was the team of friends who took over the care of our precious children during my fieldtrips to Africa—Maryrose, Breda, Patricia, Moira, Sheelagh, Mary and Fionnuala and my dear sister Joan (who even came to live in my house during one of my fieldwork trips!). I am forever indebted to you all and to all my dear friends, especially Sandra, Carmel, Sinéad, Blá, Gay, Bryan, Brian, Cormac, Noel, Marie, Ann and Adi. A special word for Dervla Murphy who has been a constant source of inspiration and support to me and to Denis Halliday for unrelenting encouragement to get this book out. Special thanks to 'my lovelies' aka my siblings, Helen, Joan and Edward, and to my dear Mum and stepfather Jim for encouraging me over that elusive finish line.

I offer my thanks to my own little family. My darling children, Caoimhe and John, you know that you are both the light of my life. Thanks for all the encouraging hugs, the cupcakes, the welcome distractions and the comic relief. To Peter, thank you for supporting me in so many ways, at so many levels over the years, including considerable technical assistance which enabled me to bring this work to completion. And to my constant four-legged companion Holly, for teaching me the value of keeping it in the moment!

CONTENTS

Abbreviations

ABN	African Biodiversity Network
ABS	Access and benefit sharing
ACB	African Centre for Biodiversity
ACC	Agricultural Commercialisation Clusters
ACTESA	Alliance for Commodity Trade in East and Southern Africa
ADLI	Agricultural Development Led Industrialisation
ADM	Archer Daniel Midland
ADP	Agricultural Development Plan (Kenya)
AFFA	Agriculture, Fisheries and Food Authority (Act)
AfrI	Action from Ireland
AFSA	Alliance for Food Sovereignty in Africa
AFSTA	Africa Seed Trade Association
AGOA	Africa Growth and Opportunity Act
AGP	Agricultural Growth Programme
AGRA	Alliance for a Green Revolution in Africa
AMDe	Agribusiness and Market Development
AoA	Agreement on Agriculture
APEC	Asia-Pacific Economic Cooperation
APHRD	Animal and Plant Health Regulatory Directorate
ARIPO	African Regional Intellectual Property Organisation
ASARECA	Association for Strengthening Agricultural Research in Eastern and Central Africa
ASCU	Agricultural Sector Co-ordination Unit
ASDS	Agricultural Sector Development Strategy
ASTA	American Seed Trade Association
ATA	Agricultural Transformation Agency
AU	African Union

BMGF	Bill and Melinda Gates Foundation
BoARD	Bureau of Agriculture Research and Development
BRICS	Brazil, Russia, India, China and South Africa
CAADP	Comprehensive Africa Agriculture Development Programme
CBD	Convention on Biological Diversity
CFS	FAO Committee on World Food Security
CGIAR	Consultative Group on International Agricultural Research
CGRFA	Commission on Genetic Resources for Food and Agriculture
ChemChina	China National Chemical Corporation
CIAT	International Centre for Tropical Agriculture
CIMMYT	International Maize and Wheat Improvement Centre
COFEK	Consumers Federation of Kenya
CoM	Council of Ministers
COMESA	Common Market for Eastern and Southern Africa
CSO	Civil society organisation
DA	Development Agent
DD	Democratic Developmentalism
DFATD	Department of Foreign Affairs, Trade and Development
DFID	Department for International Development
DUS	Distinct, uniform and stable
EAC	East African Community
EASCOM	Eastern Africa Seed Committee
EBI	Ethiopian Biodiversity Institute
EBRD	European Bank of Reconstruction and Development
ECAPAPA	Eastern and Central Africa Programme for Agricultural Policy Analysis
EG&T	Economic Growth and Transformation Office
EHPEA	Ethiopian Horticulture Producers Exporters Association
EIAR	Ethiopian Institute of Agricultural Research
EOSA	Ethiopian Organic Seed Action
EPA	Environmental Protection Agency
EPRDF	Ethiopian People's Revolutionary Democratic Front
ESA	Ethiopian Standards Agency
ESE	Ethiopian Seed Enterprise
ETC	Erosion, Technology and Concentration Group
EU	European Union
F1 Hybrid	First filial generation of bred seed from different parent line
FAO	Food and Agriculture Organisation of UN
FARA	Forum for Agricultural Research in Africa
FBSPM	Farmer-based seed production and marketing
FDI	Foreign Direct Investment
FIAN	Food First Information and Action Network

FS	Food Security
GATT	General Agreement on Tariffs and Trade
GDP	Gross domestic product
GEF	Global Environment Facility
GM	Genetically Modified
GMO	Genetically Modified Organism
GNI	Gross national income
GoE	Government of Ethiopia
GTP	Growth and Transformation Plan
HDI	Human Development Index
HoPRs	House of People's Representatives
IAASTD	International Assessment of Agricultural Knowledge, Science and Technology for Development
IBC	Institute of Biodiversity Conservation (now called EBI)
IBRD	International Bank for Reconstruction and Development
ICARDA	International Centre for Agricultural Research in the Dry Areas
ICC	International Criminal Court
ICESCR	International Covenant on Economic, Social and Cultural Rights
ICRISAT	International Crops Research Institute for Semi-Arid Tropics
ICT	Information and communications technology
IDA	International Development Association
IDLO	International Development Law Organisation
IFC	International Finance Corporation
IFPRI	International Food Policy Research Institute
IGO	Intergovernmental organisation
IMF	International Monetary Fund
INGO	International non-governmental organisation
IO	International organisation
IP	Intellectual Property
IPC	International Planning Committee
IPCC	Intergovernmental Panel on Climate Change
IPGRI	International Plant Genetic Resources Institute
IPRs	Intellectual Property Rights
IRRI	International Rice Research Institute
ISD	Institute for Sustainable Development
ISF	International Seed Federation
ISSA	Irish Seed Savers Association
ISSD	Integrated Seed Sector Development
ITPGRFA	International Treaty on Plant Genetic Resources for Food and Agriculture
KARI	Kenyan Agricultural Research Institute

KEFRA	Kenyan Food Rights Alliance
KENFAP	Kenyan National Federation of Agricultural Producers
Kephis	Kenya Plant Health Inspectorate Service
KGGCU	Kenyan Gain Growers Cooperative Union
KNHRC	Kenyan National Human Rights Commission
KSC	Kenya Seed Company
LDC	Least Developed Country
LRAN	Land Research and Action Network
LVC	La Via Campesina
MERCOSUR	Mercado Común del Sur
MLAR	Market-Led Agrarian Reform
MoA	Ministry of Agriculture
MoARD	Ministry of Agriculture and Rural Development (Ethiopia)
MoFED	Ministry of Finance and Economic Development
MoU	Memorandum of Understanding
NAFSN	North American Food Systems Network
NARI	National Agricultural Research Institute
NDA	National designated authority
NGO	Non-governmental organisation
NPGRC	The National Plant Genetic Resources Centre
NSP	National Seed Policy (Kenya)
OAPI	Organisation Africaine de la Propriété Intellectuelle
OCSIA	Office of Cyber Security and Information Assurance
ODA	Overseas development aid
ODM	Orange Democratic Movement
OECD	Organisation for Economic Co-operation and Development
OFAB	Open Forum on Agricultural Biotechnology in Africa
OPV	Open-Pollinated Variety
PA	Peasant association
PASDEP	Plan for Accelerated and Sustainable Development to End Poverty
PASS	Programme for Africa's Seed Systems (AGRA)
PBAK	Plant Breeders Association of Kenya
PBRs	Plant breeders' rights
PBS	Protection of Basic Services
PC	Primary Cooperative
PGRs	Plant Genetic Resources
PLC	Public Limited Company
PPB	Participatory plant breeding
PPP	Public-private partnership
PRONAL	Programa Nacional de Alimentación
PVP	Plant variety protection

QDS	Quality declared seed
RAFI	Rural Advancement Foundation International
RED	Rural Economic Development
SADC	South African Development Community
SHA	Self Help Africa
SNNPR	Southern Nations, Nationalities and Peoples Region
SPVAA	Seed and Plant Varieties Amendment Act
SRA	Strategy for Revitalising Agriculture
S-RWG	Seed Regional Working Group
SSASI	Sub-Saharan African Seed Initiative (World Bank)
SSCF	Seed Security Conceptual Framework
STAK	Seed Trade Association of Kenya
SWG	Seed Regional Working Group
TF1	Task Force 1
TF2	Task Force 2
TNA	Transnational actor
TNCs	Transnational corporations
TPLF	Tigrayan People's Liberation Front
TRIPS	Trade-Related Intellectual Property Rights
UNCTAD	United Nations Conference on Trade and Development
UNDAF	United Nations Development Assistance Framework
UNDP	United Nations Development Programme
UNEP	United Nations Environment Programme
UNFAO	United Nations Food and Agriculture Organisation (also FAO)
UNMDGs	United Nations Millennium Development Goals
UPOV	International Union for the Protection of New Varieties of Plants
USAID	United States Agency for International Development
USDA	United States Department of Agriculture
WFS	World Food Summit
WIPO	World Intellectual Property Organisation
WTO	World Trade Organization

LIST OF FIGURES

LIST OF TABLES

CHAPTER 1

The Core Dilemma: Seed Sovereignty and Globalisation

The central dilemma of this book is the tension between, on the one hand, the need for food, or food security, and, on the other, the desire to maintain sovereignty over food production, in this case seeds and agricultural production, or seed sovereignty. The need for food is increasingly being met by a greater reliance on uniform commercially bred seed, including genetically modified seeds designed by multinational corporations and supported by philanthropic organisations such as the Bill and Melinda Gates Foundation. However, meeting the need for food in this way risks eliminating the sovereignty of domestic producers. These are the local farmers and farming communities that wish to continue freely cultivating the seeds of their choice, notably their locally adapted seeds. The loss of seed sovereignty threatens the extinction of these varieties of seeds, ones that have been in use for millennia and that have adapted to a changing climate over that time, and central to our shared agrobiodiversity. The dilemma between food security and seed sovereignty is expressed most clearly in areas of the world where the need for food is most acute, notably parts of sub-Saharan Africa that are subject to rapid climate change. In this context, this book has two main aims. Firstly, to examine the extent to which local farmers and farming communities in sub-Saharan Africa can exercise seed sovereignty in the face of the forces of globalisation expressed, amongst other actors by multinational corporations and philanthropic organisations. Secondly, to apply existing theories of globalisation to provide a better way of understanding the contemporary exercise of seed sovereignty in this region.

© The Author(s) 2019   1
C. O'Grady Walshe, *Globalisation and Seed Sovereignty in Sub-Saharan Africa*, International Political Economy Series, https://doi.org/10.1007/978-3-030-12870-8_1

The Core Dilemma

There is undoubtedly a need for subsistence food or food security. Climate change threatens already vulnerable rain-fed subsistence farming most acutely. A 20% decrease in growing periods is projected for parts of sub-Saharan Africa (ETC 2010), with the African continent set to be hardest hit with erratic weather, decreased crop yields, crop failure, increased disease, water stress and related problems of indebtedness, aid dependency and out-migration (IPCC 2014; FAO 2011, p. 188). Estimates suggest that there will be between 40 and 170 million more undernourished people directly due to climate change, with sub-Saharan Africa faring worst (FAO 2011, p. 186).

Global agribusiness is responding to this need for food security by developing 'climate-smart' seed solutions. Building on developments in plant genetics in the twentieth century, which brought new highly bred homogenous seed varieties (F1 hybrids),[1] generally referred to as improved varieties, to fulfil the need for uniformity, productivity and the growing market for monoculture cash crops, multinational global agribusiness corporations are now set to increase the commercialisation of genetically modified (GM) seeds or transgenic seeds. By inserting particular traits, notably drought resistance or other climate-related trait from one species into another, they create a transgenic plant. They claim that these seeds have the potential to substantially increase the yields and variety of foods available globally, as well as supporting the ambitious agricultural export plans for poorer countries (Robin 2010; ETC 2010; Patel 2007).

Global agribusiness is also responding increasingly strategically to the need for food security. Multinational corporations in this domain are on target to dominate the future of seed choices, significantly changing agricultural practices worldwide. Six Western firms now control over two-thirds of the formal seed market (Oakland Institute 2017, p. 3). Prior to the acquisition of Monsanto by Bayer[2] in 2018, Monsanto was already the global market leader for vegetable seeds (Berne Declaration 2013, p. 10), while 87% of the total area devoted to genetically engineered seeds worldwide was occupied by Monsanto seeds. Other giant agri/seed corpora-

---

[1] F1 hybrid refers to filial generation 1—the first filial generation of offspring of distinctly different parental types. They are not as resilient in the second generation and so are not appropriate for seed saving.

[2] https://www.bloomberg.com/news/articles/2018-09-05/bayer-sees-earnings-lower-after-63-billion-monsanto-purchase

tions such as Du Pont, Syngenta, Badische Anilin und Soda Fabrik (BASF) and Cargill to name a few claim their role and purpose is to feed the growing world population as it ascends towards nine billion. Cargill's CEO was more forthright, calling it the 'commercialisation of photosynthesis' (Page cited in Moseley 2012).

These new seed choices come at a price though. GM seeds fundamentally threaten seed sovereignty. GM seeds are engineered and owned by corporations, with strict prohibitions on use and cannot be saved. Similarly, the more common commercially bred seeds (F1 hybrids) are not rigorous in the second generation and cannot generate robust seed that can be saved by the farmer for cultivation the next year. Instead, the farmer has to return to the company year on year and buy more seed. More than that, the contractual nature of such seed production criminalises open-pollinating seed-saving systems—the natural subsistence type of farming practiced across the world by the majority of local farmers. Sovereignty over seed choice thus moves from the local farmer/community to the corporation, indeed the multinational corporation, placing seed and food security well beyond the boundary of the farmers' fields, beyond the local community and indeed the domestic state. In this way, seed sovereignty, the "critical nexus where the contemporary battle over the means of production and consumption of food will be determined" (Kloppenburg 2010, p. 368), is inextricably linked to broader processes of globalisation.

This is an important issue. Sovereignty is a fundamental principle of politics and international relations. A vast literature examines the fate of the nation-state in light of accelerating globalisation. A multiplicity of 'new actors' now exercise power and determine change in the global order at every level (George 2015; Cerny 2009, 2010; Hettne 2009; Scholte 2008; Eriksen 2007; Held and McGrew 2007; Harvey 2003; Slaughter 2004; Hirst and Thompson 2002; Sassen 1996; Giddens 1990). In this context, the basic research question of this book is to what extent can domestic actors act independently in the face of globalisation? To what extent can they exercise sovereignty?

This question is of particular importance in those parts of the world where the domestic state is relatively weak and where global interests create the prospect of much-needed economic development. Sub-Saharan Africa is one of those regions. Africa's 'development crisis' is considered to be at the heart of a market-led globalisation, characterised by massive inequalities in power, skewed regulatory processes of state systems, economic fragility and spatial differentiation (Harrison 2010, p. 6; Maathai 2010).

In these circumstances, poor countries are mandated to liberalise their policies on trade and the free flow of capital, despite being home to some of the world's most vulnerable people, and despite the paradox of their considerable resource wealth. In short, the need for development in sub-Saharan Africa can often come at the expense of domestic decision-making sovereignty.

In the African context, I focus on food sovereignty. To what extent can domestic actors, namely governments and their agricultural agencies and institutions in Africa pursue sovereign food policies in the face of global pressures from multinational corporations, philanthrocapitalist organisations and other external actors, including dominant US interests? More specifically still, I focus on seed sovereignty. To what extent can local farmers and farming communities in sub-Saharan Africa exercise seed sovereignty in the face of global pressures towards the importation of foreign seeds, GM seeds and pressure to conform to changing agricultural practices?

From Food Security to Seed Sovereignty

There is a basic difference between food security and food sovereignty. In essence, food security means freedom from hunger. Its definition has changed over time to attempt to embrace the inherent social aspects of the concept, such as access and entitlement, and more recently nutrition. By contrast, food sovereignty in this context means the ability to take decisions about food independently and freely. This links food sovereignty fundamentally with human rights—allowing for each nation to protect and regulate the sovereignty of their domestic agricultural production, guaranteeing cultural integrity and farmers' rights.

The difference between food security and food sovereignty is important. Patel (2009, p. 665) states that "it is possible to be food secure in a prison or a dictatorship". In other words, while food security may be a necessary condition for human existence, it is not a sufficient condition for a meaningful and well-lived life. Food sovereignty is required in addition to food security. It directly extends the concept of food security to encapsulate the overtly political context of autonomy, control and power to make choices and select options within the agricultural sphere.

The concept of seed sovereignty can be understood in this context. Seed security means having enough seeds to plant a crop that will ensure freedom from hunger across the year. By contrast, seed sovereignty implies the ability to take decisions about which seeds to plant independently and freely. That is to say, the concept of seed sovereignty challenges the idea that freedom from seed want is sufficient by itself. It is possible to have seed security without exercising seed sovereignty. In this context, seed sovereignty is something to be valued in itself. It is part of a meaningful and well-lived farming life, especially in communities where there are historic cultivation practices that rely on locally adapted seeds, such as in sub-Saharan Africa.

The detailed definition of the term 'seed sovereignty' has only entered the academic lexicon in recent years, with the publication of Kloppenburg's paper titled "Seeds, Sovereignty, and the Via Campesina: Plants, Property, and the Promise of Open Source Biology" (2008), based on analysis of the historical work in the seed space, notably by La Via Campesina (LVC) and Indian non-governmental organisation (NGO), Navdanya.[3] This work establishes that seed sovereignty is associated with the autonomy and ability to save seeds, to be seed producers, to breed, share and replant seeds, and to have a juridical mandate to shape policy in that regard. It is synonymous with open-pollinated farmers' varieties (OPVs) or 'landraces' which have co-evolved with farming civilisations over millennia, adapting within and across diverse agroecologies in participatory, largely non-commoditised seed systems. It is central to the majority subsistence farming populations of the world, one billion of whom still predominantly feed the planet (GRAIN 2005, 2008). Loss of seed sovereignty is associated with loss of power and control over these precise principles and practices, regardless of what other gains might be made. Globalisation threatens the loss of seed sovereignty.

GLOBALISATION AND SEED SOVEREIGNTY

The conflict and risk which comes with a loss of seed sovereignty is immense. For example, in 2010 in Haiti, 10,000 poor people marched in the streets refusing US company Monsanto's GM seed aid, despite their dire need for assistance. They feared cross-contamination with local seed varieties and local seed production systems, as well as potential damage to local markets.

---

[3] Meaning Nine Seeds, Navdanya was established by Vandana Shiva in 1987 to protect and enhance indigenous seed and crop varieties from corporate capture in agriculture.

The food sovereignty movement has politicised the discourse around the issues of food control and governance in recent years (LVC 2008; McKeon 2015; Patel 2009; De Schutter 2009; McMichael and Schneider 2011). This activity has focussed attention back on the central issues of power, control, risks and benefits (Tansey 2011) in the seed/food political space. There have been calls for the reassertion of citizen, farmer and ecological rights for sovereign spaces of repossession and recovery (UN Human Rights Council Declaration on Peasant Rights (Art 19) and Biodiversity Rights (Art 20); FAO 2015; Kloppenburg 2014; Shiva 2013; Nagoya Protocol 2010; IAASTD 2009; De Schutter 2009; Nyéléni Declaration 2007[4]) as a counter-hegemonic force to the neo-liberal globalising effects of 'shadow sovereigns' (George 2015) in the agribusiness sector. For many of these authors and organisations, food sovereignty would also act as an essential buffer against the risks associated with narrowing the genetic resource base (Worede 2011), central to monoculture commercial farming in the face of climate change.

UN Special Rapporteur for food, Olivier de Schutter, sharpened the focus on the direct relationship between seed policies and food security especially in vulnerable communities in the Global South, when he highlighted the risk of Intellectual Property (IP) related monopoly rights neglecting poor farmers' needs, undermining traditional systems of seed saving and exchange, and "losing biodiversity to the uniformisation encouraged by the spread of commercial varieties" (De Schutter 2009, p. 2). Many authors and organisations added their voice to concerns regarding the risk attached to jeopardising farmers' seed systems (Munyi et al. 2016; Munyi and De Jonge 2015; ACB 2015; AFSA and GRAIN 2015; World Bank 2013; Louwaars et al. 2013; UN International Declaration of Peasants' Rights 2012, since passed in November 2018; Alemu 2011; McMichael and Schneider 2011; IAASTD 2009; Altieri 2009).

Many INGOs have expressed concern at recent intensification and further concentration through a concerted agenda for 'harmonisation' of seed laws and policy, particularly in the Global South (ACB 2015; GRAIN 2005, 2008; ETC 2010). The IDS Bulletin (2011) points out that African countries have been experiencing varying degrees of change in their indigenous seed systems and seed regulatory frameworks, with the advent of a plethora of new seed-related bodies and enterprises emerging and a number of significant new actors arising both internally and externally across

---

[4] https://nyeleni.org/spip.php?article290 [accessed online 10 April 2017].

the continent (IDS Bulletin 2011). Without doubt, the region has been experiencing a significant and rapid juridification in the seed space, with the potential to change agricultural practices in a profound way, with deep-seated consequences for seed sovereignty.

In this context, this research explores the extent to which states in sub-Saharan Africa can formulate sovereign domestic policy in the face of globalisation. The focus is on the passage of laws regulating the use of seeds. It addresses the motivations of national- and local-level actors as well as transnational actors (TNAs) in determining domestic seed laws in the context of intensifying global forces. When domestic decision-makers in sub-Saharan Africa have passed seed laws, to what extent have those laws protected the exercise of seed sovereignty by local farmers and farming communities?

STUDYING SEED SOVEREIGNTY: NATIONAL SEED LAWS

There is now a body of work on the issue of seed sovereignty (Shiva 2013; Kloppenburg 2008, 2010, 2014). However, there are cross-disciplinary calls for empirical studies to deepen our understanding of the policy decision-making processes and outcomes in the area of food/seed politics (De Jonge 2014; Murphy 2014; Rahmato et al. 2014; Kloppenburg 2014; IDS Bulletin 2011; Alemu 2011; Di Falco and Chavas 2009; Abay et al. 2009, 2011). This book is a response to those calls. It provides the first in-depth study of new Kenyan and Ethiopian seed laws and also provides the first local study of seed sovereignty in Ethiopia.

This book concentrates on the exercise of seed sovereignty in the period from the early 1990s to the present. Seed sovereignty has been affected by developments such as the introduction of the Agreement on Agriculture (AoA) by the World Trade Organization (WTO) in 1995, the International Union for the Protection of New Varieties of Plants (UPOV) in 1991 and the Agreement on Trade-Related Aspects of Intellectual Property Rights (TRIPS) in 2002. These developments provide an important temporal boundary for this book.

At the macro level, new seed laws have been instituted across the African continent, intensifying particularly since the 1990s. During this period, the Global South in general and Africa in particular have been recognised as a point source for North-South asymmetrical intensification in the area of plant genetic resources or seed. In other words, incursions into the seed space have intensified, with dominant industrialised countries and their seed transnational corporations (TNCs) greatly benefiting from sweeping regulations and standing to gain enormous wealth from a genetically richer South.

Africa, therefore, provides a crucially important case to test the exercise of seed sovereignty. However, Africa is a large and varied continent, stretching as it does some 11.68 million square miles and covering 20.4% of the total land area of the planet. With 54 distinct countries and 2000 languages (Mous 2003), it features a wealth of cultural and biological diversity. In this context, I chose to focus on one area, namely East Africa. Specifically, for the cross-country comparison I chose to focus on Ethiopia and Kenya.

I chose them because both Ethiopia and Kenya share certain potentially important characteristics. Agriculture is the backbone of both economies, with 83% of Ethiopians living in rural areas and engaged in agriculture (UNDP-UNDAF 2011) and 61.1% for Kenya.[5] Both countries are largely dependent on output from small-scale rain-fed farming and livestock production. In the Ethiopian context 96% of cultivated land is occupied by smallholder farmers, the national average size being as small as 1 hectare, while 43.4% of Kenyans and 30% of Ethiopians are deemed to live below the poverty line. They also share similarities in levels of debt, gross domestic product (GDP), poverty levels and overseas development aid as shown in Table 1.1.

Geographically situated in sub-Saharan Africa, they both have broadly similar climatic conditions, with this region being significantly exposed to the increasing effects of climate change and forecasts for increased food

Table 1.1 Comparison of key features in Kenya and Ethiopia

	Kenya	Ethiopia
External debt	US $16.77bn	US $17.02bn
GDP (based on PPP)	$139.4bn	$134.7bn
Value of exports	$6.27bn	$4.14bn
Agriculture as % GDP	28.9%	46%
Net ODA	$3.236bn	$3.8bn
Poverty level per head of population	45.9%	38.9%
Population less than 25 years old	61%	64%

Source: Author's collection of data from various World Bank, OECD and UNDP sources

[5] http://kenya.opendataforafrica.org/ejikndd/kenya-agriculture-sheet

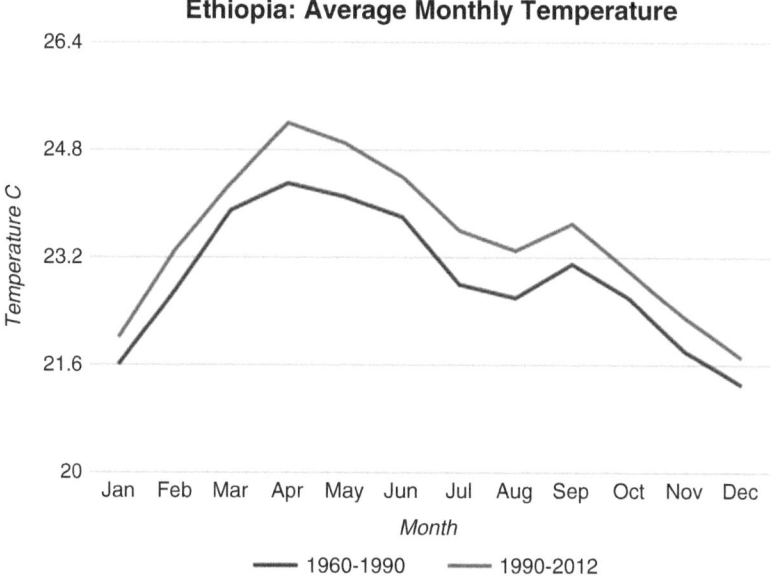

Fig. 1.1   Average monthly temperature for Ethiopia comparing 1960–1990 and 1990–2012. (Source: Adapted by the author from the original dataset produced by the Climatic Research Unit of University of East Anglia. Accessed online on 30 December 2015)

insecurity, potentially affecting up to 250 million people by 2020, with women faring worst (IPCC 2014; FAO 2011; Toulmin 2009; UNDP 2008). Figures 1.1 and 1.2 show the changing climatic conditions for temperature for both Ethiopia and Kenya for the period 1960–1990 and 1990–2012, showing the kind of marginal increases in temperature, which, coupled with more erratic rainfall, causes the kind of abiotic stress that can lead to severe environmental stresses and accompanying food crises.

In addition, both countries have a history and continuing threat of food insecurity and hunger (UNDP-UNDAF 2011; Alemu 2011; World Bank 2007), exacerbated by fluctuating climatic conditions. This makes them both susceptible to a high level of external incursions into the food/ seed space, largely through 'aid' initiatives. This aid dependency is substantial as detailed in Table 1.1, with annual overseas development aid

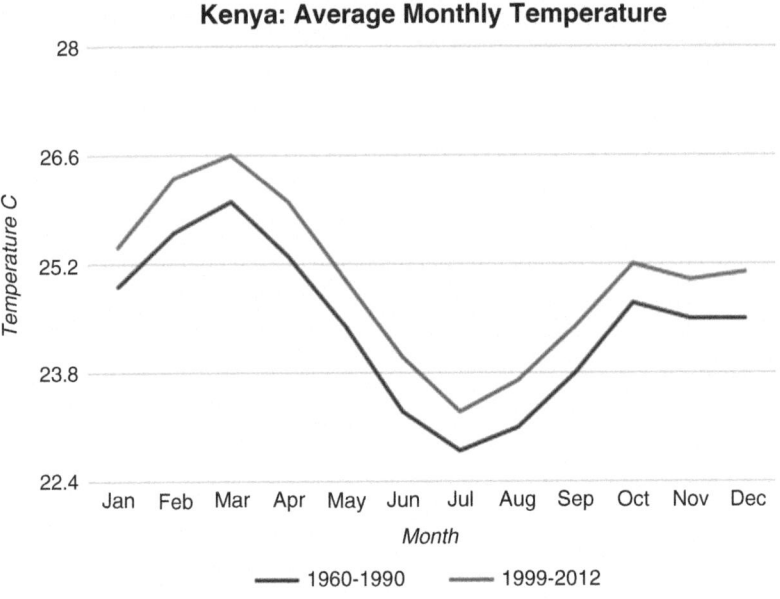

Fig. 1.2 Average monthly temperature for Kenya comparing 1960–1990 and 1990–2012. (Source: As Fig. 1.1)

(ODA) figures for Kenya and Ethiopia standing at $3.2 billion and $3.8 billion respectively (World Bank 2013).[6]

Seed aid specifically has become a notable feature of the seed systems in many African countries over many decades. For example, the United Nations Food and Agriculture Organisation (UNFAO) has implemented 400 seed relief projects in Africa between 2001 and 2003. Parts of Eastern Kenya have received seed aid almost continuously since the early 1990s, while areas of the Central and Northern Highlands of Ethiopia are reported to have received some kind of seed aid since 1974 (Sperling 2008), which is in itself a much contested aspect of the seed discourse (Scoones and Thompson 2011).

Both countries, in keeping with the 33 million small farms across sub-Saharan Africa (i.e. with less than two hectares of good arable land Scoones and Thompson 2011, p. 1), maintain vibrant informal seed economies. These farmers are presently largely reliant on Open-Pollinated Varieties of landrace (i.e.

---

[6] http://data.worldbank.org/indicator/DT.ODA.ALLD.CD

locally adapted genetically diverse) varieties of farmer-selected, produced and saved seed, evolving 'in situ', in predominantly non-commoditised agrarian systems of exchange, with figures as high as 97% for Ethiopian informal seed systems (Alemu 2011), and 78% for Kenya (Munyi and De Jonge 2015, p. 166). This is a centrally important point to this book and a fundamental issue in the literature on seed and food sovereignty, particularly regarding intellectual property laws and rights, plant-breeders' rights and the dominant narrative on promoting the formal seed system in the African context. As Scoones and Thompson outline, citing GRAIN:

> So any talk of seeds today, if it is not specifically about local or farmers' seeds, implies private seeds—seeds that farmers have to buy that come with tight restrictions on their use. (GRAIN 2008, cited in Scoones and Thompson 2011, p. 6)

Both countries have recently initiated ambitious programmes for their agricultural sectors—Ethiopia's Growth and Transformation Plan (GTP) 2011–2015 (now in stage 2—GTP2 2015–2019) and the Government of Kenya's 'Vision 30', Strategy for Revitalising Agriculture (SRA)—a ten-year action plan 2004–2014, followed by the "Agricultural Sector Development Strategy (ASDS) 2010–2020". New seed laws and regulations have become central to both countries' drive to make changes within seed practices in their agricultural sectors, part of wider plans to move to middle-income country status by 2025 in the case of Ethiopia (ATA Ethiopian Agricultural Transformation Agency 2012), and part of a market-liberalising economic agenda already spanning many decades in the case of Kenya. Ethiopia and Kenya are also identified in the literature as important areas where changes currently happening in the seed space are central to an intensifying discourse regarding power, control, risks and benefits (Tansey 2011).

I chose to study the two most recent pieces of seed legislation in the two countries, namely Kenya's Seeds and Plants Varieties (Amendment) Act (SPVAA) 2012 (Gazetted 4 January 2013)[7] and Ethiopia's Seed Proclamation 782/2013.[8] For each law, I was interested in a number of detailed questions. Who wrote each law? How was it drafted? What was the motivation behind the content of the law? Who were the key actors involved in each law? Who was included in the process of determining its

---

[7] The Seed and Plant Varieties Act, 1972 (as amended in 2002) Act No. 2 of 2002 (Cap 326).
[8] Federal Negarit Gazette, No. 27, 15 February 2013, pp. 6808–6825.

contents? Who was excluded from the process? Fundamentally, though, I was interested in one issue. To what extent did the laws protect the exercise of seed sovereignty by local farmers and farming communities? The answer to this question soon became clear. The Kenyan law did very little to protect domestic seed sovereignty. By contrast, the Ethiopian law did include measures to protect seed sovereignty. This is an important finding as it shows that the effect of globalisation on seed sovereignty is not uniform across countries, including similar countries in sub-Saharan Africa.

This finding led me to investigate the Ethiopian case in more detail.

STUDYING SEED SOVEREIGNTY: PRACTICE AT THE LOCAL LEVEL

One of the problems with comparing Ethiopia and Kenya is that even if they are similar in many respects, they also vary in terms of history and colonial legacy, which had lasting effects on agricultural systems in both countries. Ethiopia stands out in the African context because, though briefly occupied by the Italians in 1935–1941, it was never colonised, and thus was not subjected to the colonising influence of a commercialisation and modernisation programme for agriculture in the periphery to serve the needs of the core coloniser, as was the fate of other African regions. On the contrary, the Declaration of the East Africa Protectorate by the British Government in 1895 marked the colonisation of Kenya, which, alongside the construction of the East Africa railway and the Swynnerton Plan (Swynnerton 1954), is regarded as critical to changes towards commercial farming and the supporting institutional structures which followed (Munyi and de Jonge 2015, p. 163). Given the distinct historical differences, it was imperative to ensure against systematically biased results that might have been generated by comparing Africa's only non-colonised country with another country. A within-country study allows me to address that potential problem.

To examine decision-making in a more specific sub-national context, I chose to focus on Ethiopia. I chose Ethiopia because, as a recognised important primary and secondary centre of domestication for some 38 different crops, and a global centre of origin for key crops such as coffee and teff (Vavilov 1997), Ethiopia boasts 18 different major agroecological zones and 49 sub-agroecological zones, and is home to a vibrant informal seed system which is intrinsically linked to deep-rooted cultural traditions, in keeping with the majority of smallholder farmers across sub-Saharan

Africa. It is associated with a strong, centralised state structure (Alemu 2011), which extends to the seed sector in the form of a tight 'public' hold, despite government moves towards opening up the economy to some private-sector development in recent years, and differs from Kenya in this respect, as it is noted to be a more liberalised economy.

Ethiopia has maintained a vigorous policy and practice of germplasm exchange between state and the majority population of smallholder farmers, through the work of a globally renowned Ethiopian scientist at the Ministry of Agriculture from the 1970s to 1990s, Dr. Melaku Worede. As a scientist at the Ministry during the 1970s, he pioneered the establishment of a partnership between a newly formed state-run gene bank (now Ethiopian Biodiversity Institute [EBI]), where the constant exchange of germplasm in a 'conservation-through-use' programme with farmers' seed varieties, in recognition of farmers' role as the primary seed producers. This system of exchange remains the norm today. This Ethiopian model is now regarded as a 'cause célèbre' in many circles, particularly in the area of 'in situ' (on-site) 'conservation through use', participatory plant-breeding practices which maintain and enhance essential agrobiodiversity especially in climate-vulnerable zones. The Ethiopian model is inextricably linked with the definitions of seed sovereignty provided earlier. This model is also marked by very strong state involvement historically at every level of the seed sector in Ethiopia and an acknowledgement across the literature from critics and advocates alike that the Ethiopian State has historically held a tight rein on its agricultural/seed sector (Alemu 2011; Worede 2011, 1991; Feyissa 2006). Nevertheless, many changes have been afoot for a number of years, and it is clear that Ethiopia is opening up to the neo-liberal model of development at a rapid pace, enjoying double-digit growth rates, the highest in Africa according to the International Monetary Fund (IMF) (USAID country report 2011–2015), and encouraging inward investment at every level. I chose to focus on Ethiopia for my within-country study to assess these key changes happening in Ethiopia.

Within Ethiopia, I focused on an important cereal seed which provides one of the staple crops for the country, namely barley. I carried out a comparative case study of two different modes of barley production: malt barley versus food barley in a distinct location in the Highland region. This was a good choice because there are only two common types of barley grown by farmers, malt barley and food barley (Kaso and Guben 2015), each with distinct uses. Food barley is the staple crop of the poorest people in the most degraded land in the Ethiopian Highland region. By contrast, malt barley is specifically targeted by the brewing industry, including TNCs such as Diageo and Heineken.

A review of the most pertinent scientific literature as well as some preliminary interviews with key people such as barley expert Professor Fetien Abay of Mekelle University in Tigray's Crop Science Department revealed that food barley production was declining due to changing farmer choices in some areas (Abay et al. 2009, 2011). Yet, over 97% of the barley crop in Ethiopia has traditionally been produced by subsistence farmers using landraces (farmer-saved, selected and produced seed varieties) (Alemayehu 1995), which "exhibit significant genetic heterogeneity" (Di Falco and Chavas 2009, p. 599). Barley accounts for over 60% of the food consumption of the population in the Highland region (Abay et al. 2009, p. 46), where historically most of the barley varieties were food barley varieties grown to meet a myriad of culinary and on-farm needs, with less emphasis on malt barley production, which was limited to household/local beer needs (Abay et al. 2009; Di Falco and Chavas 2009). Accordingly, the national emphasis was placed on food barley seed varietal selection, enhancement, release and distribution from government research programmes in the past. This is changing now, as the emphasis has shifted from food barley seed varieties to malt barley seed (Alemu Interview 2 December 2013) as the country seeks to develop the beer sector using malt barley.

A within-country and within-region (a barley-growing region) study identifying seed choices has the potential to provide important insights as to why actors (farmers) are making the decisions they do and what forces are driving these changes in seed selection. This study met an important gap in the literature with a number of studies referred to the need to understand how policy decisions were affecting outcomes (Abay et al. 2009, 2011; Di Falco and Bezabih 2010; Di Falco and Chavas, 2009). Thus, I chose to study why farmers choose to plant food barley or malt barley.

To this end, I chose to study the regional State of Oromia in Ethiopia's Central Highlands. Barley had been traditionally grown here by smallholder farmers for subsistence, but it is now a region that has been targeted for development of malt barley production for the burgeoning brewing industry through public-private partnership with corporate drinks company Diageo and other major actors. Therefore, there has been a change in seed policy and practice on the ground. Farmers are now switching a significant portion of their land use from their own locally sourced, farmer-produced seed varieties for predominantly food barley production to externally sourced seed for malt barley production for commercial purposes, providing an important opportunity as a case study of seed sovereignty. This is a significant moment when a degree of seed sovereignty changes hands or is altered because as Worede points out "usually dependency on introduced varieties is created by the displacement of farmers' own varieties" (Worede 2011, p. 368), which he points out is happening now across Africa "influenced by modern commercial crop production systems" (ibid.).

Oromia is the largest national regional state and in terms of population size and area, representing 34.3% of Ethiopia and covering an average area of 363,375km^2. It is larger than many African countries, UK, Germany or France. Oromia regional state is classified into 18 zones, 309 districts (265 rural districts) and 44 towns, and more than 6889 peasant and urban dwellers associations (PAs).[9]

Because Oromia is so vast, it displays a wide range of features which are significant to this study, meaning that I needed to focus on one locality. Its diverse climatic condition is determined mainly by two key climatic elements, namely latitude and longitude. Despite its latitude classifying it as a tropical zone, its climate is affected by its varying altitude which in turn affects the distribution of temperature, rainfall and vegetation type, and so despite its predominantly tropical status (49.8%), Oromia has a number of other agro-climatic zones, sub-tropical (42.2%), temperate (7.5%) and Wurch[10] (0.4%), with rainfall variations from 400 mm in parts of Borena in South Oromia to over 2400 mm in parts of Illubador zone in its Western Highlands. Similarly, average temperature varies from 7.5° to greater than 22° in the Highlands and as high as 30° in the lowland region. Oromia has the highest rainfall in the country at 2400 mm per annum, which would account for its high population at 32,997,211 (Ethiopian Census 2007) and for the high level of cultivation in the region standing as it does at 28.4% of the national total. The rest of the region consists of natural forest (8.5%), shrub, bush and grasslands (59.7%), and swamps and watercourses (2.9%) (ibid.).

---

[9] Peasant associations are also called kebeles in Ethiopia or kushets in Tigray. When five or six or up to ten villages come together, they form a kebele.

[10] 'Wurch', is the climatic zone represented by cold moist temperatures associated with Afroalpine areas. The only crop capable of growing at this high altitude is barley.

All these factors were important in determining my choice of within-country, within-region case study. It was essential that the farmers in my case study came from the same locality, since farming practices vary significantly according to climate in these regions.

My within-country, within-region case study was based in the Gallema Farmer's Cooperative Union site in Arsi zone of the Oromia region. The farmers were chosen from the same local 'woreda',[11] Asella, and from the same 'kebele' or neighbourhood. They are part of the same ethnic group (in the case of Arsi, e.g. 82.93% Oromo), and though culturally mixed with 58.1% Muslim and 40.01% Ethiopian Orthodox Christians, all speak the same language (Oromo), and farm in the same climatic conditions with similar agroecological conditions. The average rural household in Arsi has 1.2 hectares of land, compared to national average of 1.01 or 1.04 for Oromia region (World Bank 2004).

The farmers in this locality are equally capable of growing both food and malt barley, and their fields lie interspersed with one and other at this location. This particular area has a history of growing local 'farmer varieties' of malt and food barley.

The key questions informing both of my within-country and within-region studies were:

1. How does the switch to malt barley production for the new value chain reflect the different accounts of globalisation?
2. What is the role of domestic actors at national and local level in determining the switch?
3. What is the role of key TNA actors in determining the switch?

The conclusion in this case is significant. The application of the value chain in this area is by no means totally globalised, nor is the state necessarily the main player, despite its central involvement. Other important actors are at the table and seed sovereignty is certainly shifting to these new actors, but in an ambiguous way, and for myriad reasons, with positive and negative flows. Seed sovereignty is still largely held in state and farmer hands, but is by no means fixed or congruent.

---

[11] A 'wereda' or woreda is a third-level administrative district division. It is part of a zone, which are grouped into regions based on ethno-linguistic communities or 'kiliochs'.

THEORISING GLOBALISATION AND SEED SOVEREIGNTY

This book does not aim to explain the variation in seed sovereignty in Kenya and Ethiopia. The reasons for the differences between the two countries lie no doubt in the differences in the political regimes of the two countries, their domestic political institutions, party politics, recent political history and many other factors. Instead, the aim is to use International Relations theory to inform how the exercise of seed sovereignty in Kenya and Ethiopia at both the national and sub-national level can be conceptualised in a globalised world. How can we best make sense of seed sovereignty in the context of globalisation? Which theories of globalisation help to inform our understanding of seed sovereignty in Kenya and Ethiopia and within Ethiopia itself?

To achieve this aim, I draw upon Held and McGrew's (Held and McGrew 2007; McGrew 2011) competing theories of globalisation, namely hyperglobalism, scepticism and transformationalism. In essence, their framework, which is directly related to how we understand the effects of multinational companies, philanthropic foundations and other global actors, is used to introduce ways of conceptualising the effect of these actors on sovereignty in a globalised world. Their framework is directly related to the central aim of this book in this regard. Adapting Held and McGrew's theoretical construct makes it possible to identify the actors involved, the coercive/persuasive nature of power at play, the winners and losers (Scoones and Thompson 2011) and the patterns of inclusion and exclusion. Crucially, their framework allows us to conceptualise the degree of agency of actors in the face of the pressures from globalisation. Central to this examination are those dominant global actors, namely, IOs, such as WTO, UPOV and the World Bank, TNAs, such as TNCs or the Bill and Melinda Gates Foundation and the INGOs, who are all charged with the erosion of the state's role under globalisation.

Set in this context, this book usefully exposes the increasingly complex, transgovernmental nature of seed sovereignty in the face of globalisation. The empirical application of Held and McGrew's framework in two countries in sub-Saharan Africa confirmed that seed sovereignty is indeed affected by global forces, but in different ways in different localities. In one case study, Kenya, the new seed law is hyperglobalised. It grants proprietorial rights over new, uniform seeds to commercial/TNC interests, dislocating a key state regulatory authority in the process and excluding civil society and farmer bodies from meaningful participation. However, extending the comparative study to Ethiopia reveals a less globalised seed law, with critical exemptions for subsistence farmers and their informal seed networks. The state still mattered, but not in a way that sceptics would suggest. Instead ambiguous, multidirectional and conflictual arrangements of shared seed sovereignty between a myriad of actors is emerging, in keeping with the transformationalist school of thought. Similarly, the within-country case study in Oromia provides another example of such ambiguity, as new highly globalised seed practices are emerging alongside clear accommodation of traditional farmer seed practices.

These case studies confirm that on balance, it is more likely that state sovereignty is neither being eroded as hyperglobalists suggest nor reasserting its power and dominance as the primary actor as proponents of the sceptical school of thought claim. Instead, as the transformationalist perspective asserts, national sovereignty remains the 'principal juridical attribute' but is increasingly divided and shared between local, national, regional and global authorities. What emerges are 'overlapping sovereignties' in complex new arrangements and heightened conflict and insecurity at all levels accompanying these new 'transgovernmental relations'. Globalising forces are transforming the state's seed policy role, opening up multiple and ambiguous pathways. The effect is potentially highly conflictual, compounding problems as the shift from government to polycentric governance in the seed area is profound.

The Role of the Researcher

The empirical work in this book is based on the analysis of documentary and archival material. However, it is also based on the many interviews that I conducted in Ethiopia and Kenya during my research trips there. In this regard, the role of the researcher is critical, particularly in an African context. I was aware of the 'insider-outsider' dilemma which is relevant to all research, and more nuanced in an African context due to historically unequal relationships between 'outsiders' and local populations, with some profound ethical implications, which cannot be ignored. The "complexity of the researcher's identity" is important in examining the layers of power at play in the research process as highlighted by Geleta (2013). Others point to the central issues of 'autonomy and agency' of both the 'researcher' and the 'researched' in the research process (Mander 2010; Pittaway et al. 2010; Smart 2009; Guillemin and Gillam 2004). Pittaway et al. highlight the importance of 'negotiating a reciprocal beneficial relationship' in what she describes as a 'new framework approach' (Pittaway et al. 2010), which resonates with calls for renewed vigour, from within sociology of 'the art of listening' to find a way of using the tools of social enquiry that give space for multiple views, without forcing coherence and logic onto the 'mess' that research can sometimes uncover (Smart 2009). This is pertinent to all research but particularly to my endeavour here. In the context of any research, acknowledging the researcher as an 'actor' in his/her own right is critical. The researcher acts as 'participant observer' but also as interpreter of information throughout the process, therefore their worldview cannot be entirely divorced from their work. For Gillham, they are "a participant observer who acknowledges (and looks out for) their role in what they discover" (Gillham 2000, p. 7). In my case, not only was I an 'outsider', European, white and female, with concomitant layers of social complexity in an African context, but I was also cognisant of my own lifelong involvement in environmental conservation work, which, in the case of this study could well be argued, carries a predilection towards the normative frame of reference, favouring indigenous seeds over corporate appropriation.

I employed a number of strategies to address these issues. I constructed my research design to tabulate and reflect the views/motivations of a diverse range of actors involved in the seed space, regardless of who they may be. In essence these opinions, facts and reflections constitute the central basis of my book, the search for the 'truth' (as contested as any truth may be), within the study, which would best be served by a robust pursuit of as much evidence as could be found, and the application of equally rigorous questioning of the diverse actors involved. As a researcher, aware of my own background in environmental conservation, I had to be conscious that the answers given were not only shaped by my presence, but could even shape their perspectives on themselves after I left. I used a number of ways to address this. Using semi-structured interviews, I subjected all interviewees/actors to the same questions regarding their role and motivations, for example their reasons for involvement or non-involvement in the seed programme, their consultation or non-consultation in relation to the seed law, constantly subjecting myself to re-evaluate any assumptions, and endeavouring to clarify and verify whenever gaps appeared.

In the context of the insider/outsider dilemma, the research is replete with national and local actors' voices throughout, no matter how contradictory or 'messy' some of these responses may be. These are the actors who are determining seed and food futures for millions of people in East Africa, and this book aims to understand what they are saying, and what the implications of their multiple actions might be.

CONCLUSION

This book presents an innovative, multilevel approach based on evidence-based, in-depth empirical case studies in two critically important countries in East Africa. It provides four important new insights. Firstly, this study applies different approaches to globalisation to the issue of seed sovereignty for the first time. Secondly, for the first time it applies these theoretical approaches to an empirical study in sub-Saharan Africa. Thirdly, this book provides the first in-depth study of new Kenyan and Ethiopian seed laws and also provides the first local study of seed sovereignty in Ethiopia. Fourthly, it provides a framework for assessing the changing nature of seed sovereignty in sub-Saharan Africa that has the potential to be applied more generally. To begin, in the next chapter I present the varied positions scholars take regarding actors ability to make independent choices in the face of globalising pressures to conform. This method of analysis provides the theoretical framework for the chapters which follow.

REFERENCES

Abay, F., Bjornstad, A. and Smale, M. 2009. Measuring on farm diversity and determinants of barley diversity in Tigray, northern Ethiopia. *Monoma Ethiopian Journal of Science*, 1(2), pp. 44–66.

Abay, F., de Boef, W. and Bjørnstad, Å. 2011. Network analysis of barley seed flows in Tigray, Ethiopia: supporting the design of strategies that contribute to on-farm management of plant genetic resources. *Plant Genetic Resources*, 9(4), pp. 495–505.

ACB. 2015. *The expansion of the commercial seed sector in sub-Saharan Africa: major players, key issues and trends* [Online]. Available from: www.acbio.org.za. [Accessed 3rd February 2016].

AFSA and GRAIN. 2015. *Land and seed laws under attack, who is pushing changes in Africa?* [Online]. Available from: http://www.grain.org [Accessed 22nd January 2016].

Alemayehu, F. 1995. *Genetic variation between and within Ethiopian barley land-races with emphasis on durable resistance*. PhD thesis. Wageningen University, Landbouw.

Alemu, D. 2011. The political economy of Ethiopian cereal seed systems: state control, market liberalisation and decentralisation. *IDS Bulletin*, 42(4), pp. 69–77.

Altieri, M.A. 2009. Agroecology, small farms and food sovereignty. *Monthly Review* [Online]. 61(3), pp. 102–113. Available from: http://search.proquest.com/openview/98165d33d3bec3c0f47ca6821e49eef0/1 [Accessed 6th December 2010].

ATA Ethiopian Agricultural Transformation Agency. 2012. *ATA Ethiopian Agricultural Transformation Agency. 2012. Program Update July 2012 / vol. 1, no. 2.* Addis Ababa: Ethiopian Agricultural Transformation Agency.

Berne Declaration. 2013. *Agropoly: a handful of corporations control world food production.* Zurich, Switzerland: Berne Declaration & EcoNexus.

Cerny, P. 2009. Multi-nodal politics: globalisation is what actors make of it. *Review of International Studies*, 35(2), pp. 421–449.

Cerny, P. 2010. *Rethinking world politics, a theory of transnational neopluralism.* Oxford: Oxford University press.

De Jonge, B. 2014. Plant variety protection in sub-Saharan Africa: balancing commercial and smallholder farmers' interests. *Journal of Politics and Law*, 7 (3), pp. 100–111.

De Schutter, O. 2009. *Seed policies and the right to food: enhancing agrobiodiversity and encouraging innovation report of the Special Rapporteur.* A/64/170 United Nations General Assembly [Online]. Available from: http://farmersrights. org/pdf/righttofood-n0942473.pdf. [Accessed 23rd February 2013].

Di Falco, S. and Bezabih, Y. 2010. Seeds for livelihood: crop biodiversity and food production in Ethiopia. *Ecological Economics*, 69, pp. 1695–1702.

Di Falco, S. and Chavas, J. 2009. On crop biodiversity, risk exposure, and food security in the highlands of Ethiopia. *American Journal of Agricultural Economics*, 91(3), pp. 599–611.

Eriksen, T.H. 2007. *Globalisation: the key concepts.* Oxford: Berg Publishers.

ETC (Erosion, Technology and Concentration Group). 2010. *Capturing climate genes. Gene giants stockpile 'climate-ready' patents* [Online]. Available from: http://www.etcgroup.org/content/gene-giants-stockpile-patents-"climate-ready"-crops-bid-become-biomassters-0. [Accessed 1st February 2012].

Ethiopian Census. 2007. *Census-2007 Report.* Addis Ababa: Central Statistical Agency. Available at: http://www.csa.gov.et/census-report/complete-report/ census-2007. Accessed 14 March 2019.

FAO. 2011. *Potential effects of climate change on crop pollination.* Rome: FAO.

FAO. 2015. *The state of food insecurity in the world.* Rome: FAO.

Feyissa, R. 2006. *Farmers' rights in Ethiopia: a case study.* Lysaker, Norway: The Fridtjof Nansen Institute.

Geleta, E.B. 2013. The politics of identity and methodology in African development ethnography. *Qualitative Research*, 14(1), pp. 131–146 [Online]. Available from: http://journals.sagepub.com/doi/abs/10.1177/1468794112468469. [Accessed 26th June 2013].

George, S. 2015. *Shadow sovereigns—how global corporations are seizing power.* Cambridge UK: Polity Press.

Giddens, A. 1990. *The consequences of modernity.* Cambridge UK: Polity Press.

Gillham, B. 2000. *Case study research methods.* London and New York: Continuum.

GRAIN. 2005. *Africa's seed laws: red carpet for the corporations* [Online]. Available from: https://www.grain.org/article/entries/540-africa-s-seeds-laws-red-carpet-for-corporations. [Accessed 11th November 2014].

GRAIN. 2008. Seed aid, agribusiness and the food crisis. Editorial. *Seedling*, October: 2–9.

Guillemin, M. and Gillam, L. 2004. Ethics, reflexivity and 'Ethically important moments' in research. *Qualitative Inquiry*, 10(2), pp. 261–280.

Harrison, G. 2010. *Neoliberal Africa: The impact of global social engineering*. London: Zed Books.

Harvey, D. 2003. *The new imperialism*. Oxford: Oxford University Press.

Held, D. and McGrew, A. 2007. *Globalisation/anti-globalisation: beyond the great divide*. 2nd edition, Cambridge: Polity Press.

Hettne, B. 2009. *Thinking about development – development matters*. London: Zed Books.

Hirst, P. and Thompson, G. 2002. The future of globalisation—cooperation and conflict. *Journal of the Nordic International Studies Association*, 37(3), pp. 247–265.

IAASTD. 2009. *Agriculture at a crossroads: a synthesis of the global and sub-global IAASTD reports*. Washington, DC: Island Press.

IDS Bulletin. 2011. *The politics of seed in Africa's green revolution*. Oxford: Wiley and Blackwell.

IPCC. 2014. *Climate change 2014: synthesis report, contribution of working groups I, II and III to the fifth assessment report of the Intergovernmental Panel on Climate Change* [Online]. Available from: http://ipcc.ch/pdf/assessment-report/ar5/syr/AR5_SYR_FINAL_SPM.pdf. [Accessed 30th March 2017].

Kaso, T. and Guben, G. 2015. Review of barley value chain management in Ethiopia. *Journal of Biology, Agriculture and Healthcare*, 5(10), pp. 84–97.

Kloppenburg, J. 2008. *Seeds, sovereignty, and the Via Campesina. Plants, Property, and the Promise of Open Source Biology* [Online]. Available from: https://www.researchgate.net/publication/255583305_Seeds_Sovereignty_and_the_Via_Campesina_Plants_Property_and_the_Promise_of_Open_Source_Biology. [Accessed 10th April 2017].

Kloppenburg, J. 2010. Impeding dispossession, enabling repossession: biological open source and the recovery of seed sovereignty. *Journal of Agrarian Change*, 10(3), pp. 367–388.

Kloppenburg, J., 2014. Re-purposing the master's tools: the open source seed initiative and the struggle for seed sovereignty. *Journal of Peasant Studies*, 41(6), pp. 1225–1246.

Louwaars, N.P., De Boef, W.S. and Edeme, J. 2013. Integrated seed sector development in Africa: A basis for seed policy and law. *Journal of Crop Improvement*. 27, pp. 186–214.

LVC. 2008. Declaration of Maputo. *IN: V International Conference of LVC, October 19–22* [Online]. Available from: http://viacampesina.org/en/index.php/our-conferences-mainmenu-28/5-maputo-2008-mainmenu-68/declarations-mainmenu-70/602-open-letter-from-maputo-v-international-conference-of-la-vcampesina

Maathai, W. 2010. *The challenge for Africa*. London: Arrow Books.

Mander, H. 2010. 'Words from the heart': Researching people's stories. *Journal of Human Rights Practice*, 2(2), pp. 252–270.

McGrew, A. 2011. Globalisation and global politics. IN: Baylis, J., Smith, S. and Owens, P. (eds.) *The globalization of world politics, an introduction to international relations*. 5th ed. New York: Oxford University Press, pp. 14–33.

McKeon, N. 2015. *Food security governance; empowering communities, regulating corporations*. London and New York: Routledge, Taylor and Francis.

McMichael, P. and Schneider, M. 2011. Food security politics and the millennium development goals. *Third World Quarterly*, 32(1), pp. 119–139.

Moseley, W.G. 2012. The corporate take-over of African food security. *Pambazuka News* [Online]. Available from: www.pambazuka.org/food-health/corporate-take-over-african-food-security. [Accessed 21st September 2013].

Mous, M. 2003. *The making of a mixed language. The case of Ma'a/Mbugu*. Amsterdam, Philadelphia: John Benjamins Publishing Company.

Munyi, P. and De Jonge, B. 2015. Seed systems support in Kenya: consideration for an integrated seed sector development approach. *Journal of Sustainable Development*, 8(2), pp. 161.

Munyi, P., De Jonge, B. and Visser, B. 2016. Opportunities and threats to harmonisation of plant breeders' rights in Africa: ARIPO and SADC. *African Journal of International and Comparative Law*, 24(1), pp. 86–104.

Murphy, S. 2014. Expanding the possibilities for a future free of hunger. *Dialogues in Human Geography* [Online], 4(2), pp. 225–228. Available from: http://journals.sagepub.com/doi/pdf/10.1177/2043820614537166. [Accessed 26th September 2014].

Nagoya Protocol. 2010. *Nagoya protocol on access to genetic resources and the fair and equitable sharing of benefits arising from their utilization to the Convention on Biological Diversity*. Quebec: Secetariat of the Convention on Biological Diversity, United Nations Environmental Programme.

Nyéléni Declaration. 2007. *Nyéléni 2007. Forum for food sovereignty. Sélingué, Mali, February 23–27, 2007.* Available at: https://nyeleni.org/DOWNLOADS/Nyelni_EN.pdf. Accessed 14 March 2019.

Oakland Institute. 2017. *Down on the seed: The World Bank enables corporate take-over of seeds*. California, USA: Oakland Institute.

Patel, R. 2007. *Stuffed and starved*. London: Portobello Books.

Patel, R. 2009. What does food sovereignty look like? *The Journal of Peasant Studies,* 36(3), pp. 663–706.

Pittaway, E., Bartolomei, L, and Hugman, R. 2010. Stop stealing our stories: The ethics of research with vulnerable groups. *Journal of Human Rights Practice*, 2(2), pp. 229–251.

Rahmato, D., Ayenew, M., Kefale, A. and Habermann, B. 2014. *Reflections on development in Ethiopia. New trends, sustainability and challenges*. Addis Ababa, Ethiopia: Forum for Social Studies.

Robin, M. 2010. *The world according to Monsanto: pollution, corruption, and the control of our food supply*. New York: The New Press.

Sassen, S. 1996. *Losing control: sovereignty in an age of globalisation.* New York: Columbia University Press.

Scholte, J.A. 2008. Defining globalisation. *The World Economy,* 31(11), pp. 1471–1502.

Scoones, I. and Thompson, J. 2011. The politics of seed in Africa's green revolution: alternative narratives and competing pathways. *IDS Bulletin,* 42(4), pp. 1–23.

Shiva, V. 2013. *The law of the seed.* [Online]. Available from: http://www.navdanya.org/attachments/Latest_Publications4.pdf. [Accessed 2nd April 2015].

Slaughter, A.M. 2004. *A new world order.* Princeton and Oxford: Princeton University Press.

Smart, C. 2009. Shifting horizons: reflections on qualitative methods. *Feminist Theory,* 10(3), pp. 295–308.

Sperling, L. 2008. *When disaster strikes: a guide for assessing seed security.* Cali: CIAT.

Swynnerton, R. J. M. 1954. *A plan to intensify the development of African agriculture in Kenya.* Nairobi: Government Printer.

Tansey, G. 2011. Whose power to control? Some reflections on seed systems and food security in a changing world. *IDS Bulletin,* 42(4), pp. 111–120.

Toulmin, C. 2009. *Climate change in Africa.* London: Zed books.

UNDP. 2008. *Human development report 2007/08. Fighting climate change: Human solidarity in a divided world* [Online]. Available from: http://hdr.undp.org/en/content/human-development-report-20078. [Accessed 10th April 2017].

UNDP-UNDAF. 2011. *Ethiopia United Nations Development Assistance Framework 2012-2015.* United Nations Country Team, March 2011.

UN International Declaration of Peasants' Rights. 2012. *Final study on the advancement of the rights of peasants and other people working in rural areas.* Geneva: United Nation A/HRC/AC/8/6. Available at: http://www.wphna.org/htdocs/downloadsfeb2013/2012%20Declaration%20of%20Peasants%27%20Rights.pdf. Accessed 14 March 2019.

Vavilov, N. 1997. *Five continents.* Rome, St Petersburg: IPGRI.

Worede, M. 1991. An Ethiopian perspective on conservation and utilisation of plant genetic resources. *IN:* Jan Engels, N.N., Hawkes, J.G. and Worede, M. (eds.) *Plant genetic resources of Ethiopia.* New York, Cambridge University Press, pp. 3–21.

Worede, M. 2011. Establishing a community seed supply system: community seed bank complexes in Africa. *IN:* Li Ching, L., Edwards, S., Scialabba, N. E. (eds.) *Climate change and food systems resilience in sub-Saharan Africa.* Rome, Italy: FAO, pp. 361–377.

World Bank. 2004. *Four Ethiopias: a regional characterization assessing Ethiopia's growth potential and development obstacles.* Washington, DC: World Bank.

World Bank. 2007. *Using value chain approaches in agribusiness and agriculture in sub-Saharan Africa: a methodological guide.* Washington, DC: World Bank.

World Bank. 2013. *Agribusiness indicators: Kenya.* Washington, DC: Agriculture and Environment Services, The World Bank.

CHAPTER 2

Understanding Sovereignty in a Globalised World

In 1921, the Irish Free State won political independence from Britain. In this sense, Ireland gained political sovereignty. Nearly 90 years later, the European Commission, the European Central Bank and the International Monetary Fund (IMF) imposed a punitive financial bailout on Ireland in the face of huge bank debts. The bailout led to a loss of fiscal sovereignty and the arrival of the 'Troika', a team of 'technocrats', from three supranational organisations who placed strict budgeting and austerity measures on the Irish economy for the next three years. "Was it for this?", read *The Irish Times* editorial of 18 November 2010, invoking Yeats and the Irish Rebellion of Easter 1916. The bailout was worldwide news, internationally humiliating and set in the context of Ireland's long and bloody history to achieve sovereignty and self-determination.

When the principle device for controlling the economy, namely fiscal policy, was taken from the Irish Government by external agencies, Ireland as an independent state actor lost its right to autonomously decide its economic programme. But is it possible for any state to exercise sovereignty in the face of global integration in the economic realm, when "the global financial flow of $2trillion per day imposes significant constraint and discipline on even the most powerful government"? (Baylis et al. 2011, p. 22). Even if it is, can sovereignty be exercised by some countries more than others and in some policy areas more than others? The concepts

© The Author(s) 2019
C. O'Grady Walshe, *Globalisation and Seed Sovereignty in Sub-Saharan Africa*, International Political Economy Series, https://doi.org/10.1007/978-3-030-12870-8_2

of sovereignty, non-intervention, territorial integrity and juridical independence have been the defining principles of modern sovereign states since the seventeenth-century Peace Treaties of Westphalia[1] and the subsequent Westphalian Constitution. At the same time, the work of Hobbes[2] and Locke[3] established the contractual nature of the state, which resolved that henceforth domestic sovereignty would be represented by the state, with government decision-making and policy centrally determined in that political space. This conception of the state and state sovereignty is what is challenged by globalisation. This book explores the extent to which domestic states can formulate sovereign domestic policy in the face of globalisation.

What is the relationship between domestic policy sovereignty and these intensifying global forces? Without doubt global institutions and new global frameworks are impacting on the core functions of states across every sphere. However, I propose that on balance, it is more likely that the policy sovereignty of the state is neither being completely eroded as hyperglobalists suggest nor reasserting its power and dominance as proponents of the sceptical school of thought claim. Rather, I suggest that the current period may be best understood from a transformationalist perspective, which argues that globalisation is a dynamic force, transforming the state's policy role, not eroding it, operating at multiple levels and in a way that is not unidirectional, that it has positive and negative flows, and is contradictory, contingent and ambiguous, resulting in a multiplicity of new, highly differentiated, conflictual and complex outcomes in different locales. There are indications of hyperglobalism, but these appear to be far from concrete and universal. The state is adapting and sharing the space of power and, therefore, sovereignty with other key transnational actors in a globalising world, leading to both integration and fragmentation simultaneously and with mixed results in different jurisdictions. Transformationalist theory provides a practical framework to address the contradictory elements of the interpenetrative processes which occur when global forces interact with domestic and local realities.

---

[1] 1648 Treaty of Westphalia followed 'Thirty Years War', which ultimately ended feudalism and gave birth to sovereign territorial statehood.

[2] Hobbes became synonymous with the realist tradition of the 'Leviathan' state, as the fear-inducing protector of the people from internal and external threat, with violence, defence and security at the heart of the newly emergent state system domestically and internationally.

[3] Locke corrected some of the asymmetrical aspects of Hobbes' concept of a one-way power system which weighted sovereign power unevenly over the subjects.

This transformationalist perspective has been best elucidated by Held and McGrew (2007) in the context of wider debates around globalisation. They specifically call for more empirical studies to ground the perspective in key locations. This is what I seek to do in this book. I aim to illuminate the perspectives on globalisation at the key sites where the processes of globalisation can be unpacked and analysed. In these situations, it becomes possible to identify the actors involved, the coercive/persuasive nature of power at play, the winners and losers and the patterns of inclusion and exclusion. Crucially, it allows us to assess the degree of agency of the actors, particularly in relation to policy sovereignty. To this end, I have adapted Held and McGrew's (2007) model as a heuristic device for the purpose of this book.

In this chapter, I begin by reviewing the literature on globalisation and sovereignty. I then identify three dominant approaches to globalisation, and review what each school of thought says about the impact of globalisation on (a) the domestic state, (b) international organisations and (c) transnational actors.

GLOBALISATION

Globalisation is a highly contested, ambiguous and contentious term. From the outset, attempts at definition highlight the considerably divergent views that exist across the literature in relation to its nature, origins, application and future direction. Many authors agree that globalisation is represented by a deepening, widening and quickening intensification of worldwide interconnectedness, a shrinking world, capable of reaching the remotest locality. Thereafter, views diverge as to whether this marks a 'new' departure in world affairs, whether this 'hollows out' the domestic state or not and whether the interconnectedness is asymmetrical or equally distributed. A central question is who or what drives globalisation? Is it 'The Washington Consensus', composed of the supranational institutions representing free-market economics, namely the IMF, World Bank and the World Trade Organization (WTO), or the powerful transnational actors (TNAs), dominated by transnational corporations (TNCs), but including a growing number of other actors of varying hues from Greenpeace to the Gates Foundation? Others point to the global issue of climate change as a rallying cry for global action, a legitimation of a need for global governance and global solutions to a global problem.

Though media theorist Marshall McLuhan (1964) first coined the phrase 'global village' in the 1960s, referring to the globally integrative effects of television, the term 'globalisation' has only come into vogue in academic circles and beyond since the 1980s. Most agree that it is the transboundary nature and 'real-time' application that are key features of this current era of globalisation (Castells 1996). The fact that a decision can be made in one place and have immediate far-reaching consequences for people thousands of miles away in another locality is central to most accounts of globalisation's manifestation, where "the sites of power and the subjects of power may be quite literally continents apart" (Held and McGrew 2007, p. 4). It is the intensification of social relations, linking distant localities in a two-way process for Giddens (1990, p. 64), a 'time-space compression', capable of condensing and merging spatial and temporal distances for geographer David Harvey (1989, 2003), the constantly interpenetrative dynamics of 'distant proximities' as elucidated by Rosenau (2003). For King and Kendall, globalisation is 'new' in that it refers to "exchanges that transcend borders, and which often occur instantaneously and electronically" (King and Kendall 2004, p. 140), thereby distinguishing it from internationalisation referring to "exchanges between nation states across borders over centuries" (ibid.).

However, globalisation is particularly associated with this rapid intensification of technological innovation. This is seen to further embed the market system as a key driver of globalisation (Hettne 2009, p. 90), and is causing many commentators to argue that globalisation is primarily an economic process (Sorensen 2004, p. 25), driven by comparative costs, technological infrastructure, economies of scale and consumer demand.

To this end, it is asserted that patterns of globalisation are driven increasingly by companies, not countries, going so far as to state that "corporations have replaced states and theocracies as the central producers and distributors of cultural globalisation" (Held and McGrew 2007, p. 39). The impact is profound in terms of cultural flows in every direction, but also in financial flows and loss of sovereignty. McGrew highlights the fact that:

> Transnational corporations (TNCs) now account for between 25–33% of world output, 70% of world trade, (the exchange of goods and services) and 80% of international investment, while overseas production by these firms considerably exceeds the level of world exports, making them key players/ drivers in the global economy controlling the location and distribution of economic and technological resource. (McGrew 2011, p. 16)

The increased concentration of power and wealth of corporations in the twentieth century is deeply transformative. According to Willets, TNCs now number 82,100 with 807,400 foreign affiliates to these parent companies (Willetts 2011, p. 328), an increase from 2005 figures tabulating TNC numbers at 77,000 with 770,000 foreign subsidiaries selling €22.17 trillion of goods and services within every continent, equivalent to some 50% of world gross domestic product (GDP) and employing 62 million workers (UNCTAD 2006, p. 18).

Giddens asserts that "[i]f nation states are the principle 'actors' within the global political order, corporations are the dominant agents within the world economy" (Giddens 1991, pp. 70–71).

But TNCs are not the only new TNAs now crowding the stage. The TNA space is shared by other powerful players, ranging from environmental and human rights non-governmental organisations (NGOs) to international drug cartels and terrorist organisations. There are 7600 international non-governmental organisations (INGOs), such as Amnesty International and Greenpeace, 240 intergovernmental organisations (IGOs) such as the UN, the WTO, the World Bank as well as an unprecedented number of civil society organisations and many other less well-established networks of NGOs and international caucuses (Willetts 2011, p. 328), all exercising power and influence within countries and between countries, but importantly many of these bodies are not answerable to the state nor do they derive their legitimacy from traditional realms of power and authority.

Globalisation scholars draw particular attention to the emergence of the powerful International Organisations (IOs), the supranational financial institutions, namely, the World Bank, the IMF and the WTO. These IOs are now recognised as key drivers of globalisation, all of which promote trade and financial liberalisation. The establishment of the WTO in particular is recognised as the major landmark achievement of globalisation, especially the Trade-Related Aspects of Intellectual Property Rights (TRIPS) and accompanying Intellectual Property Rights (IPRs), which were central precursors to the establishment of the WTO in 1995 (Rangnekar 2014; Murphy 2013; Tansey 2011; De Schutter 2009; Downes 2003; Shiva 2001). According to Downes, TRIPS was included on the WTO agenda after "intensive lobbying by a small coterie of thirteen multinational corporations who comprised the Intellectual Property Committee (IPC)" (Downes 2003, p. 1).

Many authors highlight the domination of US hegemony and economic liberalism as a key driver, which some suggest is now, or potentially could give rise to globalisation's demise for varying reasons which I return to below (Rosenberg 2005; Hirst and Thompson 2002). For Chomsky, it is US hegemony which caused the collapse of Bretton Woods in 1971, when the "US suspended convertibility from dollars to gold … whereby the US dollar became the 'reserve currency' for the other countries within Bretton Woods" (Chomsky 2008). Chomsky specifically highlights this as an important moment where in effect globalisation was "appropriated by the powerful and became a process whereby international economic integration solely privileges the rights of investors and lenders" (ibid.). This view is echoed by Harrison and others who assess the asymmetrical effects of globalisation in Africa and the Global South, seeing "capital and increasing concentration and mobility as the key drivers to an extensive and protracted neoliberal social engineering on the continent" (Harrison 2010, p. 5), bringing with it an increased uncertainty about the whole project (ibid., p. 147).

Even politics itself is being globalised according to former US President Bill Clinton, signalling a shift from big government to global governance of what are considered to be global problems. This is the central plank of the work of development theorist Bjorn Hettne. Hettne identifies three key areas which have global implications for world political order—climate change, global civil war and the international financial crisis (Hettne 2009, p. 104). For Hettne, these represent profound implications for world political order in terms of (a) structure, which he defines as the distribution of power and resources; (b) mode of governance, seen as the avenues of influence on decision-making and policy-making; and (c) the form of legitimisation, which he sees as the basis on which the system is made acceptable to the constituent units (Hettne 2009, p. 20).

Hettne attests that globalisation is giving rise to a "post-national logic", nurturing the "association of a transnational assumption of responsibility" (ibid., p. 22), and with considerable implications for security, order and territorial sovereignty, echoing Giddens' assertion that "a post-traditional social order is emerging as a direct result of globalisation" (Giddens 1994, p. 5).

Against this backdrop, the question is whether the intensification, new multi-layered governance structures and thickening transnational interconnectedness are causing greater fragmentation or deeper integration? Is it eroding the ability of the state and other actors to act independently or not? What is the effect of globalisation on domestic policy sovereignty? In this next section, I will look at how people have tried to capture the general relationship between the two.

GLOBALISATION AND SOVEREIGNTY

To what extent can domestic actors exercise policy sovereignty in the face of globalisation? Various writers have tried to summarise the different positions in the globalisation debate over the past number of decades. From Giddens (1990) and Held's (1995) early work, which attempted an analysis of different waves or dimensions of globalisation, to Sklair's (2002) identification of key research clusters on globalisation, and later Held and McGrew (2007) and Hettne (2009), authors have attempted to differentiate and theorise the term. Many positions emanate from the competing schools of thought on world politics and international relations, the intellectual precursors to emergent approaches to globalisation. These range from neo-liberalism and realism of varying hues to Marxism, world systems theory, dependency theory, development studies and later to the constructivist, post-structuralist, post-modern and post-colonial schools of thought, before the more recent approaches to globalisation began to emerge as distinct perspectives.

Earlier theories continue to provide a focus for international relations' scholars and are clearly relevant to the issues of sovereignty and globalisation. Indeed, I am going to look at the sceptical school, with historical roots to realism in more depth below. However, because I am specifically addressing the effect of global forces on policy sovereignty in different localities, I wish to operationalise those distinct perspectives from globalisation theories, some of which draw upon the vast historic scholarship, but which provide a closer fit for empirical examination on the ground. My focus is on laws and policies and the impact of globalisation's reach on autonomous decision-making in domestic settings. In this context I am pursuing those authors who have specifically attempted to capture the general relationship between globalisation and policy sovereignty.

Here, I focus on Held and McGrew's (2007) book *Globalisation/Anti-Globalisation, Beyond the Great Divide*. This is the first detailed academic work which organises the scholarship on globalisation. They identify different approaches to globalisation (ibid., p. 5). This differentiation acts as a simplifying device to interpret the complexity of the forces shaping the contemporary world. Using this framework, it becomes possible to see the

Sceptics   Transformationalists   Hyperglobalists

Fig. 2.1 The globalisation spectrum. (Source: Author's adaptation of Held and McGrew 2007, p. 5)

varying assumptions of those working on the issue of globalisation and sovereignty.

Here, I adapt Held and McGrew (2007, p. 5) heuristic device and simplify it, making a linear spectrum and identifying hyperglobalist, sceptical and transformationalist positions (see Fig. 2.1).

Following Held and McGrew, and based on the three main approaches to globalisation in Fig. 2.1, I adapt three different interpretations of globalisation and policy sovereignty. They can be categorised as follows:

a) Hyperglobalists who primarily regard globalisation as a distinct 'new' phenomenon which is having considerable effect, particularly on state sovereignty, with a significantly increased role for international organisations and transnational actors in key policy arenas.
b) Sceptics who regard globalisation as not 'new', arguing that the state and its sovereignty remain central despite globalising forces.
c) Transformationalists who take a middle view arguing that there are varying effects of globalisation on states exercising policy sovereignty.

This differentiation is a useful analytical heuristic device, which makes it easier to access the varied positions within the literature on globalisation, and greatly assists the comparative method. It is neither conclusive nor exhaustive. It should also be immediately acknowledged that some authors overlap between the different positions on some issues and in some circumstances.

As noted above, central to the debate about whether the state is being eroded or not is the increasing role of other players in global political arenas, namely IOs, such as the World Bank or the WTO, and TNAs, such as TNCs or NGOs. I will examine each of the three perspectives on globalisation, namely hyperglobalist, sceptical and transformationalist, with regard to these three aspects, namely (1) the domestic state, (2) IOs and (3) TNAs. For each of the three basic positions, I identify the different stances taken by authors on each of these three issues (see Table 2.1).

Table 2.1 Competing approaches to globalisation

	State	*IOs*	*TNAs*
Hyperglobalists	Deterritorialised	IOs set worldwide rules	They operate worldwide
Sceptics	State-centred	IOs operate by state-to-state bargaining	Vehicles for state interests?
Transformationalists	Some states are stronger than others—US hegemony, Chinese in Africa	In IOs some states are more important than others	Some are more dominant than others

Hyperglobalists

The hyperglobalists argue that globalisation is a profoundly new era, which is characterised by denationalisation and deterritorialisation, and which is resulting in the deconstruction of the sovereign nation-state across all key areas of policy. They range between what Held and McGrew call 'globophobes', who are characterised with a condition termed 'globo-phobia', to 'globophiles' categorised as 'globophilia' (Held and McGrew 2007, p. 2). Globophobes assert that globalisation is devouring the policy sovereignty of states and oppose it, while globophiles agree that policy sovereignty is being eroded but delight in the resulting liberation from the overly bureaucratised state institutions into a borderless free world. Therefore, whilst hyperglobalists may argue from the position that globalisation is the new over-arching power house, they have very different normative positions and come from varied theoretical and methodological traditions, as well as different political persuasions, ranging as they do from the extreme left to the extreme right of the spectrum. However the hyperglobalists recognise this era as distinctly different in terms of how globalisation is organised and managed, with particular reference to the new institutional forces of globalisation, which they regard as limiting state power, causing the erosion of the core state functions and the collapse of state sovereignty. They concentrate on the evolution of key individual agencies, new non-governmental authorities/actors, such as the WTO, which is hollowing out the Westphalian system of state sovereignty and international order and the decision-making that determined it.

For the hyperglobalists, globalisation's reach is increasing in velocity and depth with a multiplicity of new governmental and non-governmental transnational actors. Hyperglobalists examine this reach through economics, referencing the emergence of new and varied global actors across domains of influence where once the state held a monopoly of power and control. Leading thinkers in this area outline the new and emerging conceptualisations of 'space', 'distance' and territory as central to a changing global polity and society.

The Domestic State

For Strange (1996), state fragmentation amounts to a retreat of core state functions as key areas that once defined the very idea and raison d'être of the state are being eroded, with the contingent movement of power and resources increasingly shifting elsewhere. Ten key areas are identified where state dominance is under threat, including its historic and monopolistic control in defence against violence, in maintaining the value of the currency, in provision for the welfare of disadvantaged citizens, in responsibility for taxation, in control of foreign trade and in the choice of an appropriate form of capitalist development all of which are diminishing from external forces (Strange 1996, p. 190). But what or who is causing the erosion into all these key areas of the domestic state? Geographer David Harvey (1989, p. 240), whom I place in the sceptical school, first coined the phrase 'time-space compression' to attempt to capture 'the shrinking world' which was evolving. This temporal and spatial flattening or deterritorialisation is regarded by the leading hyperglobalist authors (Ohmae 1995; Scholte 2008) as the central dynamic of globalisation.

The 'abolition of every possible remoteness' was first referenced by Martin Heidegger, as he assessed the impact of technological advancements, particularly media/television (Heidegger 1971, p. 165). The "deterritorialisation" (Scholte 2008, p. 1479) resulting from this alteration of social space, which "puts any person into relation with the entire world" (ibid.) is caused by what Scholte terms the "spread of transplanetary or supraterritorial relations between people", and marks a striking break from the "territorialist geography that came before", which has been a defining feature of the sovereign domestic state since Hobbes. But for Scholte also, "place is no longer territorially fixed" (ibid.). This 'new borderless transworld' is best illustrated with the example of the New York doctors performing the first transoceanic robot-assisted telesurgery by Internet on a patient in Strasbourg in 2001 (Scholte 2008, p. 1486). For Scholte this new phenomenon of technological and social organisational change is evident everywhere now, from cyberspace to global ecology with planet-wide climatic change, species loss and genetic engineering to global travel, global military, global consciousness, global law, global production and consumption, global communications and global money and finance, all central drivers of a rapid expansion and intensification of economic and legal globalisation.

For hyperglobalists, political changes, particularly post-Cold War, together with changes in the structure of the international economy and technological advancements, all forced a realignment of relations between states and redefinitions of statehood itself. The new logic of an expansionary global market is central to much hyperglobalist thinking, drawing on very different perspectives.

For Japanese business strategist Kenichi Ohmae, a recognised 'globophile' hyperglobalist, the most profound change is in the sphere of economic globalisation and the accompanying neo-liberal world order. This, for Ohmae, marks a distinct new epoch in human history, where "traditional nation states are becoming unnatural, even impossible business units in a global economy ... where economic power and political power are becoming effectively denationalised and diffused" (Ohmae 1995, p. 5, p. 149). For Ohmae, "nation states are dinosaurs waiting to die" with both state and government unnecessary middlemen, a "paralysing force a lot of the time to global solutions" (ibid., p. 4). Economic globalisation is a new current "owing nothing to the lines of demarcation of traditional political maps ... nation states have already lost their role as meaningful units of participation in the global economy of today's borderless world" (ibid., p. 11), and he predicts that states that do not adapt to this new reality will "stagnate and erode opportunities for their people, as investment and information just gets diverted elsewhere" (ibid., p. 12). This borderless world however is what puts neo-liberals at the heart of the hyperglobalist project, as Held and McGrew recognise that this is what neo-liberals attest is the great possibility that hyperglobalisation can provide "human freedom and prosperity unencumbered by the dictates of a stifling public bureaucracy and the power politics of states" (Held and McGrew 2007, p. 189). To this extent hyperglobalists see globalisation's reach into core policy areas, where states once led as the most effective means of entry and a mechanism for change to global rules.

IOs

During the anti-capitalist demonstrations in Genoa in 2001, Hardt and Negri wrote an insightful opinion piece for *The New York Times*, which highlights core elements of hyperglobalist thinking on IOs. They state:

The protests had to be directed at international and supranational organisations, such as the WTO, IMF, World Bank and G8, based on the recognition that no national power is in control of the present global order. (Hardt and Negri 2001)

For Hardt and Negri, we are in a new era of historically unprecedented global domination, with globalisation at its centre. It is a new era of 'Empire'. They describe it thus:

Empire establishes no territorial centre of power and does not rely on fixed boundaries or barriers. It is a decentred and deterritorialising apparatus of rule that progressively incorporates the entire global realm within its open, expanding frontiers. (ibid.)

In this hyperglobalist worldview, the supposed demise of the domestic state is attributed to the reputed sharing of power and authority with new loci, particularly these key IOs, new formal supraterritorial organisations and institutions with transglobal networks that increasingly determine the rules of the game. Their increasing incursions particularly into economic and legal affairs, as well as social, cultural and political realms, formerly the sole domain of discrete state networks, is a subject which preoccupies hyperglobalist scholarship.

Baldwin states that it is "Difficult to sustain any effective trade bloc when part of your wall encompasses the enemy camp" (Baldwin quoted in Held and McGrew 2007, p. 81), echoing Rosenau's assertion that globalisation is increasing the "porosity of domestic-foreign boundaries" (Rosenau 2000, p. 1). This is resulting in a "levelling [of] the laws and legal systems" to accompany and consolidate key aspects of global trade, finance and market liberalisation (Wang 2004, p. 480).

Wang points out that "the WTO now regulates almost every aspect of the world economy" (ibid., p. 479) and is now the defining institution which "effectively constitutes the normative and legal foundations of global markets and their operation" (Held and McGrew 2007, p. 81). This specific WTO case of 'institutionalisation' is critically important in the context of hyperglobalist scholarship, because it is regarded as a

seminal development in the political construction of a truly global trade system … it is not simply about trends in world trade but the critical importance of global and transnational trade authorities in the constitution of global markets. (ibid.)

Most importantly, the globophobes point out that these processes of globalisation in the making of what Hardt and Negri call "Empire", are considered "enduring rather than contingent" (Held and McGrew 2007, p. 171).

Many hyperglobalists identify this juridification surrounding the establishment of the WTO as an inherent part of the globalisation of law that is occurring with the homogenising of laws and legal systems to fit their rules (George 2015; Rangnekar 2014; Wang 2004; Gill 2003; Hardt and Negri 2000). These three IOs, alongside the G7 and Russia (formerly part of the G8), the European Union (EU) (as well as Asia-Pacific Economic Cooperation [APEC] and Mercado Común del Sur [MERCOSUR]) (Held and McGrew 2007, p. 23) are what globophobe hyperglobalists see as a new neo-liberal 'cosmocracy', which subordinates the world's poor to the vested interests of powerful elites (ibid., p. 137), and which ultimately serves only global corporate capitalism. For Gill (2003), the intensification and extension of capitalist globalisation means that "greater aspects of human activity and life forms are now subject to exploitation and commodification" (Gill 2003, p. xv). Gill describes it as a revolutionary process leading to the "creation of a world market … where dominant power and propertied and globalising elites, such as the World Economic Forum are effectively creating a new world order, comprising ideas, institutions and processes in movement" (Gill 2003, p. xvii).

Whilst the perspectives on economic globalisation and its impact range between Marxist analysis (Harrison 2010; Gill 2003; Sklair 2002), Empire (Hardt and Negri 2000) and economic liberalism (Ohmae 1995), hyperglobalists, whether globophobes or globophiles, agree that the global market economy has created a borderless world, where new transnational and supranational institutions, dominated by the financial IOs are at the centre of the globalised project, albeit in a new deterritorialised world, where the state has been swept away. Critically for these scholars, this era of globalisation is unprecedented historically, omnipresent and enduring.

Transnational Actors
Hyperglobalists point to the significance that with globalisation, territorial borders no longer demarcate the boundaries of national economic or political space, thereby breaking down the age-old distinction between internal/domestic and external space. Where states once led, now TNAs of every hue reign. These TNAs now act and increasingly dominate in every arena according to hyperglobalists, from technics (IT and

Communications) to all aspects of capital markets, environmental degradation, the migration of peoples as well as 'the new global division of labour driven by multinational corporations' (Held and McGrew 2007, p. 9). However, the role of these transnational actors is considered by hyperglobalists to be new, significant in its impact and enduring as traditional arrangements of power and legitimacy increasingly shift beyond the state frontier.

The rise of TNCs as the dominant transnational actors wielding their corporate power is one of the areas that hyperglobalists identify as a key driver of new systems of production, finance and consumption, with today's patterns of globalisation driven by companies rather than countries (ibid., p. 39). Seen as a form of 'private tyranny', for George they are *Shadow Sovereigns*:

> It's not just their size, their enormous wealth and their assets that make TNCs dangerous to democracy. It's also their concentration and cohesion, their cooperation and capacity to influence, infiltrate and in some areas virtually replace government. (George 2015, p. 18)

Their role in uneven development and their ungovernability is a cornerstone of globophobic hyperglobalist critique. They argue that the TNCs cannot be considered to be national corporations, even if they happen to be American or Japanese. According to Sklair, "the corporations see themselves as globalising, as TNCs now have more economic power at their disposal than the majority of the countries in the world, a fact borne out by annual Fortune 500 listings" (Sklair 2002, p. 36). Patel (2007) points out that "it has led to a world which is dominated by a few corporate buyers and sellers" and the ensuing colossal market power to a decreasing number of giant corporations. Peter Willetts (2011, p. 329) points to recent United Nations Conference on Trade and Development (UNCTAD) figures which state that:

> In 2008, the 50 largest transnational industrial companies, by global scales, each had annual revenues greater than the GNP of 125 members of the United Nations.

Berry asserts that these TNCs have "obtained the natural rights of citizens without assuming responsibility in proportion to their influence on public welfare" (Berry 1999, p. 119), echoing what the voices of the critical globalisation school say is part of a new era of empire—the emergence of a historically unique form of global domination with globalisation at its core (Hardt and Negri 2000). George points to the business elites and their new Davos-inspired Global Redesign Initiative, who have set their sights on managing worldwide public policy, intent on "replacing old, worn-out government multilateralism with an entirely new concept of global governance" (George 2015, p. 135).

For George this is "a new ideology of selfishness, greed and cruelty—the Great Neo-liberal Regression, where TNCs exercise illegitimate power, eroding democratically elected government in the process" (ibid.). This has profound implications for weaker/poorer states, where the asymmetrical nature of globalisation is felt most acutely. Castells (1996) is deeply critical of the 'Washington Consensus' arguing against the globophile account, which favours deregulation, removal of barriers and less government interference as proffered by noted globophile, Kenichi Ohame (1995). Globophobe hyperglobalists point to the deepening divide between rich and poor in an increasingly polarised world. In this scenario Intellectual Property Rights (Tansey 2011; ETC 2009; Robin 2010; Downes 2003; Shiva 2001) and investor rights dominate economic concern, while the interests of people are 'incidental' (Chomsky 2008).

McMichael and Schneider highlight this in the context of the controversial issue of land grabbing, stating that it is "sponsored by organisations such as the World Bank, its International Finance Corporation (IFC), the International Rice Research Institute of the CGIAR, and the European Bank for Reconstruction and Development" (McMichael and Schneider 2011, p. 123). They point also to new TNAs such as the Bill and Melinda Gates Foundation (BMGF) and Alliance for a Green Revolution in Africa (AGRA), philanthrocapitalist organisations, which operate freely in the new globalised neo-liberal borderless world, answerable and accountable to no one.

In summary, hyperglobalists see the state being 'hollowed out' under the pressure of new global forces, with profound consequences for the global polity. For globophiles this is considered advancement. The slow-moving, overly bureaucratic state is an inefficient mechanism for business. Further liberalisation for globophiles can take us to new frontiers faster under the inherent dynamic of the new borderless globalised world. For globophobic hyperglobalists this erosion of power and accountability at local, regional and national levels is impoverishing the many for the sake of the few, largely global business elites. The asymmetrical application of globalisation for hyperglobalists is the clearest example of how the increasing intensification and concentration of TNC and IO supranational transworld power is eroding systems of accountable state sovereignty, with dire consequences for humanity.

Sceptics

Hyperglobalists emphasised the deterritorialising nature of 'time-space compression', the erosion of traditional boundaries, the sweeping away of protectionist bureaucracies, the erosion of state and government control and the opening up of societies to vast real-time influences and impacts across the economic, social, cultural and political spheres, particularly in key areas of policy. But the extent to which these developments are truly eroding the power of the domestic state to exercise sovereignty is challenged by the sceptics. Once again it is important to point out that there is a wide net of methodological and normative difference on the sceptical spectrum. This is an intellectual terrain that stretches from traditional Marxism as espoused by scholars such as Harvey and Rosenberg to structural realists like Krasner and Haass, and in between, leading thinkers such as Hirst and Thompson, Giplin, Robert Keohane and Stiglitz. For some, 'globalisation' is just a buzz word that lacks any profound theoretical foundation (Rosenberg 2000). Rosenberg calls for the need to identify the "causes of globalisation and the agents behind them", rather than globalisation being seen as a cause in itself or inevitable (Rosenberg 2000, p. 1). For Rosenberg globalisation is 'epiphenomenal' "analytically redundant … bad empirics and bad theory" (ibid.). What they have in common, though, is that they see the state in whatever configuration as the primary actor and central fulcrum of power-determining outcomes both in its internal affairs and externally in its relationship to other states. In this section, I will address the sceptical position looking at the domestic state, the role of international organisations and the role of transnational actors.

The Domestic State

The sceptics do not deny the facts of globalisation, with its increased velocity (Gilpin 2001), interconnectedness (Hirst and Thompson 2002; Keohane and Nye 1977; Keohane 2001) and dependency between economies and societies (Haass 2005; Keohane 2001; Gilpin 2001), but they interpret these developments in a way that is consistent with older schools of realism, the dominant theory of world politics with a lineage dating back to

Hobbes. They argue that the world is still divided into sovereign territorial nation-states, or 'discrete competitive national states' with powerful institutions of state, whose arms extend over all matters from cradle to grave of the citizenry. This central functional distinction and political purpose facilitates the retention of the role of nation-states as key actors upholding policy sovereignty and acting rationally. The state is the central fulcrum of power, selecting strategies to maximise benefits and minimise losses, maintaining their status as the principal agents in shaping world order, where inequality, conflict, the power of the nation-state and competing political actors dominate and determine political agendas (Krasner 2001, 2005). They argue that it is not the velocity, but the impact of globalisation that matters. It is how states are making or not making decisions that is important (Gilpin 2001, p. 354).

They point to the 202 sovereign states, doubling since 1945 and including 192 members of the UN arguing that globalisation does not render obsolete the struggle for political power between those states, nor does it transcend the international political system of states that was first envisaged by Hobbes and perfected by Locke and Rousseau and the subsequent historical transformations which came to embody this system to the present day.

A leading proponent of the structural realist perspective, Kenneth Waltz, emphasises the self-interested nature of states in the international system (Waltz 1979). Sceptical scholars see the state as 'egoistic value maximisers' (Lamy 2011, p. 124), asserting that "ultimately we all still look to the state to solve the problems we face, and the state still has a monopoly over the legal use of coercive power" (ibid., p. 93), echoing Max Weber's original definition as elucidated in his 1918 paper *Politik als Beruf* that the state holds the monopoly of the legitimate use of physical force within a given territory. Krasner similarly looks to these features of the state, particularly the strong state and its increased capacity for "regulation, surveillance and extraction of resources" (Krasner 2001, p. 314). Stiglitz famously says that "Globalisation has been oversold" (Stiglitz 2005, p. 229). Others say regionalisation and internationalisation as well as intensifying geopolitics among dominant nation-states are shaping world politics rather than globalisation (Gilpin 2001). This era is seen by sceptics as "the era of the state" with the "state as the main actor and sovereignty its distinguishing trait" (Dunne and Schmidt 2011, p. 93). Keohane and Nye (1977, pp. 392–395) point out that increased global interconnectedness is causing a concomitant decline in the scope of strategic policy choices available to individual governments and the effectiveness of many traditional policy choices becoming redundant. However, Keohane, whilst acknowledging globalisation's effect on state operations, supports the Hobbesian proposition of building cooperative structures and institutions of interdependence and governance "essential if people are to have opportunities to pursue the good life" (Keohane 2001, p. 1). Hay and Marsh also make the case for state interventionism, arguing that it has been effective against the idea that globalisation is undermining the nation-state (Hay and Marsh 2000). This kind of 'statism', it is argued, was successful in East Asia, where a three-pronged system of restriction on cross-border capital flows, limitation on foreign competition and a nurturing of domestic industry underlined state power and achievement of state goals despite globalisation, and showing signs of 'an older style mercantilism' (Held and McGrew 2007, p. 198).

So to sceptics, "states still matter", but more than that, they highlight that "national or local factors, from resource endowments to state capacity, are perhaps of increasing significance in lifting nations and communities out of poverty" (Held and McGrew 2007, p. 122). Krasner (2001) and Gilpin (2001) highlight the exaggeration of globalisation's effect in this regard, saying that it "blinds scholars to how states continue to use their power to implement policies to channel economic forces in ways favourable to their own national interests and ... favourable share of the gains from international economic activities" (Gilpin 2001, p. 21).

For sceptics, globalisation is reinforcing the role of major states as 'international actors' and by inference, sovereignty remains a central part of the suite of choices in the domestic and foreign policy portfolio to be used where appropriate or beneficial to the power elites within dominant states.

IOs

Hyperglobalists pointed out that the increasing power and influence of IOs was diminishing the role of the nation-state in a profound way. However, leading authors from the sceptical school, such as Hirst and Thompson (2002), specifically highlight the period 1850–1914, arguing that it was much more intensely globalised and globalising than the present period. For them, many pronouncements on globalisation lack historical rigour and analysis in understanding evolutionary societal processes at work.

For many sceptics, this is a core point. They do not argue that globalisation is not a reality, but argue cogently that it is nothing in comparison to that period of the mid-nineteenth century which saw an explosion in communications, invention, trade and global movement with the industrial revolution and colonial expansion, the rise of corporations, a period of significantly greater interdependence than now "technological change in the form of international telegraph cables and unified markets which led to price and interest rate convergence of a kind that has never been equalled since" (Hirst and Thompson 2002, p. 248). However, US hegemony is widely regarded by sceptics as being central to understanding how power still operates at the geopolitical level despite globalisation, and how IOs are used to achieve the goals of dominant states and consolidate their hegemony. Sceptics point to the 'rational egoist' nature of states, such as the USA to convey advantage to the hegemonic power, particularly in their dealings with both formal international organisations and other transnational actors, be they TNCs or NGOs (Harvey 1989, 2003; Krasner 2001, 2005; King and Kendall 2004).

Some of the sceptical scholarship on the 1994 WTO Agreement highlights a key difference with the hyperglobalist perspective.

> There is no doubt that the eventual agreement in the WTO on intellectual property in 1994 (tied importantly to trade) reflected the interests of the major US corporations, who have a significant comparative human capital advantage in their ownership of abstract objects. (King and Kendall 2004, p. 170)

Thus, many sceptics regard the control of IOs by the dominant states, notably the USA, as central to their thesis that geopolitics and strong state power politics remain centre stage. The USA as the "prime information economy and the main net exporter of intellectual property" was an example of how coercion and bilateral trade-offs were used to achieve "wider multilateral agreements" at a later stage from key countries (King and Kendall 2004, p. 170). Sceptics argue that this talk of "globalisation" is really the "continued development of the international system of commercial liberalism" (Hirst and Thompson 2002, pp. 472–496), and thus nothing new at all.

Rosenberg asserts that because of this continued power structure of vested interest and inequality, globalisation is just a conceptual 'folly'

(Rosenberg 2005, p. 14). Rosenberg argues that globalisation was a fad of the 1990s and that normality has been restored with geopolitical manoeuvring again returned to big power politics, unilateralism post 9/11, failure to negotiate climate talks, the crippling of the International Criminal Court (ICC) and the "multiple crises of the international organizations (UN, NATO, EU) in the run-up to the second Iraq war" (Rosenberg 2005, p. 3). Others identify these institutions as prime examples of geopolitical manoeuvring; with strong states opting in and out of key global decisions at will in everything from economics to military security to climate change (Willetts 2011). This has currency with many sceptical authors, who argue that the majority of international economic transactions and political activity is concentrated within the Organisation for Economic Co-operation and Development (OECD) states (Held and McGrew 2007; Gilpin 2001; Harvey 1989) and is by no means truly global, with poorer countries significantly weakened in the present scenario where economic liberalisation is not matched with international systems of equality and solidarity (Maathai 2010; Cheru 2002). Martell asserts that the global economy is "internationalised and triadic rather than global and its internationalisation is not unprecedented" (Martell 2007, p. 175). Others identify this specifically with 'The Americanisation of World Order' (Harrison 2010; Held and McGrew 2007). It is well captured by Rafael and Stokes in their essay on the US energy and oil nexus in West Africa. They state:

> The US framework since the post-war era has been based on an open-door trading regime underpinned by American military might, positioning itself as global hegemonic stabiliser while fostering forms of geopolitical and economic dependence on US power itself. (Rafael and Stokes 2011, p. 915)

While Rafael and Stokes are looking at US interests and big power play in the energy field in a 'new scramble for Africa', the point applies for US interests on the continent in general. This is highlighted by many authors and largely referred to as 'The Washington Consensus', first mooted by John Williamson. For the sceptics, this power structure underlies the reality of who controls the power to govern the international economy, with many authors asserting that it is likely to remain in the hands of the wealthy nations, and the supranational bodies that they control and fund, like the IMF, the World Bank, the WTO and the strong states, such as the USA (Rafael and Stokes 2011; Hirst and Thompson 2002; Krasner 2005, 2001).

Tim Stevens holds that "cyberspace, independent of the extant system of sovereign states remains a distant dream" (Stevens 2013, online). He asserts that we are witnessing the state striking back in many countries, particularly in the wake of WikiLeaks and cites structural changes and legal interventions such as recent agreements like the Council of Europe's Convention on Cyber-crime (2001), to the UK establishment of the Office of Cyber Security and Information Assurance (OCSIA), as well as the development of "cyber weapons ... that can attack and compromise foreign assets with potentially remarkable accuracy" as examples (Stevens 2013, online).

The overriding and common assertion of sceptics in these scenarios is that the state remains the primary actor and all institutional appendages, old and new, ultimately succumb to state power and authority.

Transnational Actors

As already noted, sceptics point to US hegemony and the 'rational egoist' nature of dominant states to serve their own interests as central to understanding the geopolitical power relations that are occurring between major global actors, be they states, formal international organisations or other transnational actors, ranging as they do from TNCs to NGOs (King and Kendall 2004, pp. 169, 170). Krasner points out for example that "for all of the talk of growing NGO influence, their power to affect a country's domestic affairs has been limited compared to governments, international organisations, and multinational corporations" (Krasner 2001, p. 26).

There are obvious examples of the increased intensity and velocity of globalisation, and shifting boundaries and borders of social and political geography, but many sceptics assert that one of the most endemic features of globalisation is the inequality, exclusion and uneven power which exists between states, global governance favouring global capitalism and the technocratic nature of decision-making which excludes those with a legitimate stake in the outcomes (Harvey 2005). Similarly, notwithstanding the obvious increase and penetration of NGOs of varying hues onto the global stage particularly since the 1980s, many sceptics still hold that 'states remain special players' (Sorensen 2004, p. 70). They set the rules 'other actors' play by, including most of the NGOs and civil society organisations, which many sceptical authors identify as Western or Northern with the concomitant bias, but also whose lobbying efforts are mainly directed at states and their policies (Clark 2011, p. 553).

In the case of TNCs, Sklair makes a distinction between what he acknowledges as their gigantic force, but asserts that "few if any are actually global yet … free of the state and inter-state system" (Sklair 2002, p. 35). But he does point out that these TNCs and the "transnational capitalist class that owns and controls them all over the world ensures the hegemony of capitalist globalisation in the present era" (ibid.). This underlies the core of the sceptical position regarding transnational corporations as key transnational actors, which are often used by hyperglobalists as the example of how globalisation is eroding everything. Sceptics assert that TNCs play a critical role in the continued dominance of strong states (particularly the USA) and their economies. Willetts points out that

> in 2008, among the 100 TNCs with the highest level of assets outside their own country, 57 were from twelve Western European countries, 19 from USA, four with dual headquarters in Western countries, ten from Japan, three from South Korea and one each from Australia, Canada, China, Malaya, Mexico, Singapore and Hong Kong. (Willetts quoted in Baylis et al. 2011, p. 330)

Hirst and Thompson, in a prescient article, highlight some of the serious threats to the future of globalisation, most notably the paralysing chaos which climatic change will wreak, particularly in the 'developing world' but obviously not confined to it (Hirst and Thompson 2002, p. 250), where the continuation of power politics renders transnational actors less powerful than dominant states, highlighting the continued military and economic dominance of the West, notably the USA and its allies. Hirst and Thompson argue that these powerful states will still determine outcomes, including inaction, even in the face of mounting tensions, increased migration, population and deprivation wreaked by climate/environmental devastation (Hirst and Thompson 2002, p. 251). Similarly, they foresee conflict over the legitimacy deficit of WTO, IMF and World Bank, not just by China, Russia and other rising states, but also by NGO coalitions and protest groups, which, they predict, will undermine globalisation and the possibility of 'cosmopolitan democracy' as events unfold (ibid.).

In summary, sceptics accept the basic concept that there is increased velocity and a time-space compression in this era. But they dismiss claims that this tendency is "new" and that it is undermining state sovereignty, ultimately asserting that "globalisation is what states make of it". Sceptics, representing varying normative positions, still see the state as the central fulcrum of power, authority and legitimacy, dominating and determining outcomes internally and externally, across all policy areas, but most notably maintaining their control over the use and means of violence/security. States act as 'rational egoistic maximisers' for state benefit at all times. Sharing policy sovereignty has been part of the real-politick of 'organised hypocrisy' since the inception of the state system, an inherent part of the geopolitical arrangements of power and domination. To this end, IOs are directed by strong states and used to achieve the goals of dominant states and consolidate their hegemony. Few corporations are truly global according to the sceptics; rather they act to ensure capitalist globalisation on behalf of dominant states. Other TNAs, from INGOs to cyberspace are all grounded in a locality and never beyond the control of state action, whether in their lobbying efforts or the restrictions placed on them by state regulation. Globalisation is a passing fad. The state remains the primary architect and actor, constantly adapting to new geopolitical arrangements of power, consolidating its hegemonic position in decision-making and core policy arenas among other actors.

Transformationalists

As the name suggests, this group of authors see certain transformative aspects emerging from the current era of globalisation. They argue that it is creating new economic, political and social circumstances, which though uneven, are certainly transforming the state and its key areas of policy sovereignty as we have known it. Transformationalists can often seem like hyperglobalists because they see the present era as essentially globalising, and accept many of the patterns and drivers identified by leading hyperglobalists. They agree that we live in an increasingly borderless world, where 'time-space compression', particularly in this new era of technics, communications, environmental degradation and 'multi-layered governance' is central to changes occurring, shifting the foundations of legitimacy, authority and sovereignty in the process. However, they also emphasise the capacity of the state to readjust and adapt itself to new power structures, the central tenet of the sceptical school. Specifically, they recognise that globalisation is affecting different policy areas in different ways, is contingent and is replete with contradictions and ambiguities (Martell 2007, p. 186). They pay particular attention to the role of transnational actors, be they supraterritorial international organisations or bodies, and international non-governmental organisations, be they corporations, business interests or civil society, arguing that globalisation is 'amenable to political reform' (Held and McGrew 2007, p. 163), but also that it is multidirectional, with flows of ideas, people, information, capital and goods, giving rise to multiple possible futures, including increased conflict, chaos and the possibility of 'cosmopolitan democracy'.

The Domestic State

Transformationalists see the state as neither eroded nor omnipotent. They hold that states and societies are undergoing "profound transformations as they adapt to a globalising world—the globalised condition" (Hettne 2009, p. 89), resulting in a varying sovereignty or asymmetrical sovereignty across states, blurred boundaries between domestic and international sovereignty, and differentiated consequences. Held and McGrew believe that both the hyperglobalists and the sceptics exaggerate the role of globalisation in deflating or inflating the role of state sovereignty, causing misinterpretations of contemporary politics (Held and McGrew 2007, Chap. 11). For Held, globalisation has been creating new economic, political and social circumstances, which, however uneven they may be, are transforming state powers (Held 1995). For Martell, this makes globalisation an open-ended and uncertain venture (Martell 2007, p. 176), with globalisation causing a shift in the geography of political power and political organisation. Held and McGrew explain:

> For the transformationalists, it is this apparent dislocation or destabilising of the institutional coordinates of social life that is the source of both heightened conflict and insecurity at all levels, from the local to the global. (Held and McGrew 2007, p. 169)

This has considerable implications for national sovereignty, which whilst remaining a "principal juridical attribute" (McGrew 2011, p. 29), is increasingly divided and shared between local, national, regional and global authorities.

The recognition of this extension of the social, economic and political space under globalisation is a central hallmark of transformationalist scholarship, and they point to new sources of 'governance', not controlled by states in the globalised era, thus differing from the sceptics, who saw the state as dominant in international cooperation and the hyperglobalists who saw the state as in retreat (Sorensen 2004, p. 59). These "transgovernmental relations" and the new and dense web of policy networks between countries is, according to Slaughter, "rapidly becoming the most widespread and effective mode of international governance" (Slaughter 1997, p. 185), and evidence for transformationalists that a profound restructuring is taking place, with a seismic "shift from government to governance", in many ways compounding problems on a global scale from finance to ecological in its wake (Held and McGrew 2007, p. 170). Slaughter's (2004) thesis points to a 'new world order', where there is an increasing role for global public policy networks with burgeoning transnational legal and political relationships and directives, all of which are superseding domestic governance structures, causing a fragmentation at the domestic state level. Nevertheless, she asserts, there is still an important role best played by the domestic state in this new web of power, a key feature of transformationalist thinking, which espouses the multidirectional, contradictory nature of globalisation, as it binds and fragments structures simultaneously.

Leading transformationalists, Held and McGrew build on this open-ended and contradictory nature of globalisation, arguing that there is no simple logic to global integration and convergence, making the case for transformationalist thinking against the simplification of hyperglobalist and sceptical viewpoints (Held and McGrew 2007, p. 169). They regard globalisation as "dialectical, integrating and fragmenting, uniting and dividing the world by creating winners and losers, and including and excluding locales as it proceeds" from economics to cultural, social and political dimensions (ibid., p. 169). Similarly, Eriksen asserts that globalisation entails both processes of homogenisation and heterogenisation (Eriksen 2007, p. 12), that though driven by economic and technological advancements, it creates a system that is multidimensional. It is not a "unidirectional process" (ibid., p. 9). For Eriksen, because globalisation does not have a prescriptive mission per se, it does not necessarily entail the production of homogeneity, but is more inclined towards "organising heterogeneity". He argues that the local continues to thrive despite degrees of standardisation, yet the local is increasingly "enmeshed in transnational processes"—what he and others refer to as "glocalisation" or "the insertion of the local society into globalisation" (Hettne 2009, p. 87). These transformations are happening all the time through deepening relations internationally among states, transnationally, among regions and localities and translocally across social, economic, cultural and political realms. These transformations are seen to be altering the context of the state's existence and operations (Cerny 2009, p. 423). In this way many transformationalists are exercised about how the rules of sovereignty are being transformed, not eroding sovereignty per se, but states actively bargaining within it.

Cerny, who straddles the three schools in his thinking, sees the possibility of rethinking politics, arising from such multiple equilibria emerging from the present course. He sees the role of 'actors' and their agency in the 'field of action', that is, the nation-state and the state system, as the main field of potentiality, where the interests and values of different groups or 'actors' can change the substance of political outcomes in a dynamic process of what he terms 'structuration'. This, he describes as being:

> the interaction of structural constraints and the strategic and tactical choices of 'actors', which leads to wider systemic outcomes a form of pluralism or neo-pluralism ... which can be transformative/emancipatory on the one hand or remain structure bound on the other. (Cerny 2009, p. 423)

This echoes some of what Held and McGrew hold as central to the transformationalist course now where:

> Pluralisation of political orientations and allegiances can be linked to the erosion of the state's capacity to sustain a singular political identity in the face of globalisation. (Held and McGrew 2007, p. 178)

Similarly, when Rosenau refers to the increased "porosity of domestic-foreign boundaries" (Rosenau 2000, p. 1), accompanied by the dynamics of a 'fragmegrative' (ibid., p. 4) globalisation, that is, the simultaneous integration and fragmentation associated with globalisation, he highlights the encompassing tensions between national and transnational systems, between core and periphery (Rosenau 2003), as consensus erodes with an increasingly diversifying and fragmenting society. A 'new space of flows' has come to exist beside the 'old space of places' (Castells 1996, p. 348), as territorial integrity, a fundamental building block of the domestic state is swept aside. In this scenario, states remain important, though increasingly become more "polymorphous entities, diffused into complex networks involving a range of other actors" (Sorensen 2004, p. 35). This increased complexity and disaggregation of authority and diffusion of legitimacy is intensified by the inherent dynamic of the deep drivers of globalisation as referenced earlier. According to the transformationalist school, this move away from state-centric political hegemony towards emphasising partnerships between governmental, paragovernmental and non-governmental organisations, in which the state is only the 'first among equals' (Jessop 1997, p. 574) is the central hallmark of the era of globalisation and the decline of the state as core duties and functions are increasingly derogated to significant others.

IOs

Transformationalists can be distinguished from other schools of thought by not imbuing globalisation with any particular 'telos', according to Held and McGrew. Globalisation is not necessarily going to lead either to anarchy on the one hand, or global peace and prosperity on the other (Held and McGrew 2007, p. 170). Because transformationalists see globalisation as such an open-ended 'work in progress', there is a strong normative thrust in the literature, their thesis being that "Globalisation can be better and more fairly governed, regulated and shaped" (Held and McGrew 2007, p. 194), making "a second Great Transformation possible" (Hettne 2009, p. 89), as they look to methods and strategies of rethinking traditional organising assumptions and institutions of modern political life, with a particular concern for moving from sovereignty to 'cosmopolitan democracy' (Held 1995).

Transformationalists assert that there are deep structural consequences to the intensification of transworld cooperation, wrought by the increased influence of key international organisations, like the WTO, which has power of sanction, or the International Criminal Court (ICC), which has global jurisdiction over people, for example. This can "multiply the complexity of modern societies and thereby their governance, while simultaneously creating a range of new transnational problems from global warming to global financial instability, which are hugely difficult to resolve" (Held and McGrew 2007, p. 170). Transformationalists differ from hyperglobalists and sceptics in their assertion that this increased international cooperation offers new opportunities for regulation as well as simultaneously putting new constraints on states. States are becoming stronger in some aspects and weaker in others (Sorensen 2004, p. 71). This is significant, as this opening up of core state functions will not have an evenly distributed effect—weaker states being more susceptible than stronger ones to exploitation.

In this context, transformationalists point to a whole new set of transboundary problems arising due to the decisions made in a distant location (be it a country, a corporation or an IO), which can have immediate and considerable impact on people in other societies far away, given the increased velocity and mobility of globalising forces. This is resulting in increasing dislocation and destabilisation, the kind of 'multiple equilibria' which Cerny (2009) posits "where territory, identity, economy, sovereignty and the state itself no longer appear historically fixed and congruent" (Held and McGrew 2007, p. 169).

This fluidity and dislocation is highlighted by transformationalists in looking at "the distributional consequences" of globalisation. Cohen believes that in the era of globality, "states, once totally dominant in their own territories, has now become something like that of competing firms in an oligopolistic industry" (Cohen 2007, p. 214), where "money's deterritorialisation alters the structure of governance in global currency relations" (ibid., p. 208). However, he says more emphatically that deterritorialisation "has not totally deprived states of their capacity to act on behalf of their citizens. But it does oblige states now to share authority with key market agents, each side playing a critical endogenous role in an ongoing dialectical process" (ibid.), thus emphasising the changed landscape which has seen increasingly powerful private non-state actors taking over space formerly occupied by state controls and governance. In this way, the state is forced to share authority with non-state market agents, each side playing a key role in an ongoing dialectical process. Cohen (2007) asserts that, while most countries will continue to have a preference for maintaining traditional monetary sovereignty, empirical evidence would point to the fact that the extent of the ability to follow will be greatly determined by the size of the country and the extent of its political and economic linkages, as well as domestic politics that the country enjoys. This echoes the concern over the uneven nature of global institutional involvement and its varied effect on actor sovereignty, particularly states, and highlights the 'skewed regulative processes, economic fragility and spatial differentiation', which accompanies much economic globalisation (Harrison 2010, p. 6; Maathai 2010).

For transformationalists, this potential crisis of legitimacy in global governance, and the distributional consequences of a polarising globalisation, which increasingly marginalises the majority poor, holds the seed of a counter-hegemonic drive as those disaffected by IOs and other transnational actors use the tools of globalisation to transform and reform these institutions and forces (Evans 2008; Hettne 2009; Held and McGrew 2007; Scholte 2008). Yet in other cases these institutions themselves are claiming radical changes are afoot. For example, former WTO chief, Pascal

Lamy, points out that the "developing countries GDP surpassed the developed world in 2012 for the first time in centuries" (*The Irish Times* 19 April 2013, p. 5), claiming changing patterns of investment and industrialisation in recent years, when after a brief downturn post-9/11 figures for global trade and investment have recovered and surpassed previous levels, testament to what transformationalist authors assert is the possibility of reforming globalisation.

But this will not be easy, as Keohane (2001, p. 2) predicts that "at the global scale, the supply of rogues may be expected to expand with the extent of the market", arguing for greater institutional protection and security from the arbitrary nature of other actors at national and importantly at a global level, while others see this as potentially undermining the pursuit and potential realisation of a genuine 'cosmopolitan democracy' (Held and McGrew 2007). Hettne identifies this as a "governance gap", where, he says, "Economics has become global, but politics is still largely national" (Hettne 2009, p. 116). He calls for "the possibility of a rules-bound order, a refutation of the anarchy model of international relations as well as the utopia of the self-regulating market", which he says is possible in formulating a new "great compromise" and "framework for global development ... to deal with the disrupting social consequences of asymmetrical and polarised deterritorialisation under market-led globalisation" (Hettne, ibid.). This cosmopolitan normative thrust of the transformationalist analysis is a distinct characteristic of this school of thought, which argues "for ethical and humane globalisation, combining economic efficiency with equity or social justice ... nothing less than a reform and transformation of the existing infrastructures of globalisation, with those international organisations and institutions as key targets for reform and transformation now" (Held and McGrew 2007; Scholte 2008).

Transnational Actors

Scholte argues that this large-scale transplanetary, supraterritorial connectivity, though dominated by more powerful actors, nevertheless gives an opportunity to many social actors "admittedly unequally to respond and to mould the trends" (Scholte 2008). For him, this at least provides a potential disruptor effect with both 'enabling and disabling' potential in local and global spheres (ibid.).

Many transformationalist authors point to inequality and exclusion as one of the most endemic features of globalisation—uneven power between states, global governance favouring global capitalism and the technocratic nature of decision-making which excludes those with a legitimate stake in the outcomes. However, simultaneously, this inequality and exclusion underlines the increased state of flux wrought by globalisation. The ensuing tension is seen as a critical moment, as alternative avenues of transformation are emerging (Held and McGrew 2007), in what Cerny sees as

> a pluralistic constellation of actors operating across increasingly diverse, 'multinucleated' transnational spaces opening up a range of alternative outcomes and multiple equilibria … a new complex phenomenon of 'multinodal politics'. (Cerny 2009, p. 449)

This is a core point for transformationalists, who look at key drivers, such as technics, favoured by hyperglobalists and see transformative effects. Because transformationalist thinking does not imbue globalisation with any particular goal or purpose as mentioned earlier, they look to the positive and negative flows, without getting caught up in hyperglobalism or bogged down in state-centrism. For example, they point to the Information and communications technology (ICT) sector and accept that it can be used by transnational corporations for empire building, for homogenising cultures and disabling economies, but also point to the mobile-banking revolution which has happened in Africa in the past decade, for example, where people who never had a bank account have leap-frogged directly to an innovative mobile-banking revolution (James and Versteeg 2007) ahead of the rest of the world, often charged by solar mobile-charging stations in remotest areas, and reflecting the possible cultural, social, economic and political possibilities from such contingent aspects of 'glocalisation' (Hettne 2009). Half of the one billion people on the African continent now have a mobile phone, when there was only four million phones in 1998 (Carmody 2011; James and Versteeg 2007). Sassen points to the untapped potential of such a phenomenon in her assessment of the use of social media in the Arab Spring movement (Sassen 2011). This multidirectional possibility of many aspects of globalisation is a key aspect of transformationalist thinking.

Others point to remittances, which are worth 1% of sub-Saharan Africa GDP, 4% in Middle East and North Africa, and migration, which has both positive and negative dimensions (Carmody 2011).

For transformationalists they accept the hyperglobalist position that these other supraterritorial TNAs are now exercising considerable power, influence and change 'transworld' with an unprecedented velocity and depth. Whether these are drug cartels, TNCs such as Du Pont, Monsanto, Shell Oil or Coca Cola, or INGOs such as Greenpeace, Amnesty International or the World Wildlife Fund, terrorist groups, private individuals or capitalist philanthropists, they are all transnationalising, non-state global actors. Because of this independence of the baggage of state/institutional bureaucracy, they have varying levels of accountability to traditional institutional power systems, deriving their legitimacy and authority outside of such systems, thereby throwing up other possible outcomes for an emerging global polity. This autonomy/separateness may be perceived as bad in terms of increased insecurity as drug cartels and terrorist groups increase numbers and power, or good if there is increased environmental protection or changed environmental policy, when groups like Greenpeace are successful in stopping the dumping of oil platforms in the North Sea, achieving global bans on landmines or stopping transboundary pollution with the successful passage of the Basel Convention. Or culturally, it manifests itself through what Tucker (1997, p. 17) calls cultural "hybridity", or a "global melange", found, for example, in the merging of Irish music with the music of nomadic bands from Mali.

Evans (2008) directly addresses this issue, asserting that actor potential for transformation is possible now like never before, precisely because of the intensification of corporate capitalist concentration and the self-regulation of the market which has created the perfect conditions for a counter-hegemonic globalisation, echoing some of the sceptical predictions of Hirst and Thompson (2002) earlier. This counter-hegemonic movement according to Evans is perfectly placed to organise itself globally, appropriating the "tools and resources of generic globalisation" to do so, specifically referring to the need to protect society and nature on the one hand and the potential chaos of an unsustainable market on the other (Evans 2008, p. 275). This is echoed by Scholte when he identified the "unequal opportunities" (Scholte 2008, p. 1497), which the disruption wrought by globalisation brings to social actors, but which enables social movements to "respond to and mould this trend" (Scholte 2008, p. 1498).

So, for transformationalists, the field of action lies in that place where the contradictory forces of globalisation are most evident, where the variable geometry and asymmetrical nature of globalisation is most acutely felt. For transformationalists, globalisation needs to be empirically grounded in the localities where 'distant proximities' have become a reality, where a multiplicity of actors are forced to contend with the structural consequences of globalisation's reach and the concomitant increased societal complexity.

In summary, transformationalist theorists argue that we are in new and unchartered waters. The time-space compression accompanying this era of globalisation is creating multiple crises and complex opportunities, where a multiplicity of new actors now share the policy space where the state once led, creating blurred boundaries between domestic and international sovereignty and with very differentiated consequences from the local to the global. Despite these new complex arrangements and because globalisation lacks any specific telos, transformationalists insist that states are now disaggregated players, but far from eroded. National sovereignty remains the 'principal juridical attribute', but key constituent parts/policies and core functions are increasingly shared between local, national regional and global authorities and institutions. It is a multidirectional process, not unidirectional. Globalisation, for the transformationalist school is contingent, ambiguous and contradictory, making it both a danger and an opportunity, constantly integrating and fragmenting under different circumstances in different locations. To this end, many transformationalists pursue a normative avenue, seeking to reform the excesses of globalisation, cognisant of the uneven application of the distribution of its benefits, and insisting that by its nature, it holds the possibility for counter-hegemonic possibilities at every differentiated turn.

CONCLUSION

This chapter has examined how globalisation has challenged the traditional state-centric concept of territorial sovereignty. It presented three different positions which seek to understand where sovereignty now lies in the face of such global forces. I adapted Held and McGrew's (2007) work, which identified three distinct schools of thought or competing perspectives, hyperglobalist, sceptical and transformationalist, and reviewed how writers in each school perceive the relative impact of global forces on policy sovereignty and the ability of actors, be they domestic states, IOs or other transnational actors, to act independently. In the next chapter, I narrow the focus of enquiry from globalisation's effect on sovereignty generally to specifically examining how the different interpretations can be used to examine the issue of seed sovereignty.

REFERENCES

Baylis, J., Smith, S. and Owens, P. 2011. *The globalization of world politics, an introduction to international relations.* 5th ed. New York: Oxford University Press.

Berry, T. 1999. *The great work, our way into the future.* New York: Bell Tower.

Carmody, P. 2011. *The new scramble for Africa.* Cambridge: Polity.

Castells, M. 1996. *The rise of the network society: the information age: economy, society and culture.* Vol. 1. Oxford: Blackwell.

Cerny, P. 2009. Multi-nodal politics: globalisation is what actors make of it. *Review of International Studies,* 35(2) pp. 421–449.

Cheru, F. 2002. *African renaissance: roadmaps to the challenge of globalisation.* London: Zed Books.

Chomsky, N. 2008. Anti-democratic nature of US capitalism is being exposed. *The Irish Times* [Online], 10th October. Available from: https://www.irishtimes.com/opinion/anti-democratic-nature-of-us-capitalism-is-being-exposed-1.894183 [Accessed 13th October 2012].

Clark, I. 2011. Globalization and the post-cold war order. *IN:* Baylis, J., Smith, S. and Owens, P. (eds.) *The globalization of world politics, an introduction to international relations.* 5th ed. New York: Oxford University Press, pp. 544–558.

Cohen, B.J. 2007. *Global monetary governance.* London and New York: Routledge, Taylor and Francis.

De Schutter, O. 2009. *Seed policies and the right to food: enhancing agrobiodiversity and encouraging innovation report of the Special Rapporteur.* A/64/170 United Nations General Assembly [Online]. Available from: http://farmersrights.org/pdf/righttofood-n0942473.pdf [Accessed 23rd February 2013].

Downes, G. 2003. *Implications of TRIPS for food security in the majority world* [Online]. Available from: http://comhlamh.org/wp-content/uploads/2013/09/Implications-of-Trips-for-Food-Security.pdf [Accessed 10th April 2017].

Dunne, T. and Schmidt, B.C. 2011. Realism. *IN:* Baylis, J., Smith, S. and Owens, P. (eds.) *The globalization of world politics, an introduction to international relations.* 5th ed. New York: Oxford University Press, pp. 84–99.

Eriksen, T.H. 2007. *Globalisation: the key concepts.* Oxford: Berg Publishers.

ETC (Erosion, Technology and Concentration Group). 2009. *Who will feed us? Questions for the food and climate crises* [Online]. Available from: http://www.etcgroup.org/sites/www.etcgroup.org/files/web_who_will_feed_us_with_notes_0.pdf [Accessed 1st February 2012].

Evans, P. 2008. Is an alternative globalisation possible? *Politics and Society,* 36(271), pp. 271–305.

George, S. 2015. *Shadow sovereigns—how global corporations are seizing power.* Cambridge UK: Polity Press.

Giddens, A. 1990. *The consequences of modernity.* Cambridge UK: Polity Press.

Giddens, A. 1991. *Modernity and Self-Identity: Self and Society in the Late Modern Age.* Cambridge: Polity Press.

Giddens, A. 1994. *Beyond left and right: the future of radical politics.* Cambridge: Polity.

Gill, S. 2003. *Power and resistance in the new world order.* Basingstoke: Palgrave.

Gilpin, R. 2001. *Global political economy.* Princeton University Press: Princeton.

Haass, R.N. 2005. Sovereignty, *Foreign Policy,* no. 150, pp. 54–55.

Hardt, M. and Negri, A. 2000. *Empire.* Cambridge: Harvard University Press.

Hardt, M. and Negri, A. 2001. What the protestors in Genoa want. *New York Times,* 20th July 2001 [Online]. Available from: www.nytimes.com/2001/07/20/opinion/what-the-protesters-in-genoa-want.html. [Accessed 2nd December 2016].

Harrison, G. 2010. *Neoliberal Africa: The impact of global social engineering.* London Zed Books.

Harvey, D. 1989. *The condition of postmodernity.* Oxford: Blackwell.

Harvey, D. 2003. *The new imperialism.* Oxford: Oxford University Press.

Harvey, D. 2005. *A brief history of neoliberalism.* Oxford: Oxford University Press.

Hay, C. and Marsh, D. 2000. *Demystifying globalization.* London and New York: Macmillan Press.

Heidegger, M. 1971. *"The thing" in poetry, language, thought.* New York: Harper and Row.

Held, D. 1995. *Democracy and the Global Order: From the Modern State to Global Governance,* Cambridge: Polity Press.

Held, D. and McGrew, A. 2007. *Globalisation/anti-globalisation: beyond the great divide.* 2nd edition, Cambridge: Polity Press.

Hettne, B. 2009. *Thinking about development—development matters.* London: Zed Books.

Hirst, P. and Thompson, G. 2002. The future of globalisation—cooperation and conflict. *Journal of the Nordic International Studies Association,* 37(3), pp. 247–265.

Irish Times. 2010. Was it for this? (Editorial). *The Irish Times* [Online]. Available from: https://www.irishtimes.com/opinion/was-it-for-this-1.678424. 18th November 2010. [Accessed 17th October 2015].

James, J. and Versteeg. 2007. Mobile phones in Africa: how much do we really know? *Social Indicator Research,* 84, pp. 117–126.

Jessop, B. 1997. Capitalism and its future: remarks on regulation, government and governance. *Review of International Political Economy,* 4(3), pp. 561–581.

Keohane, R. 2001. Governance in a partially globalised world: presidential address. American Political Science Association 2000. *American Political Science Review,* 95(1), pp. 1–13.

Keohane, R.O. and Nye, J. 1977. *Power and interdependence.* Boston: Little Brown.

King, R. and Kendall, G. 2004. *The state, democracy and globalisation.* Hampshire and New York: Palgrave Macmillan.

Krasner, S.D. 2001. Sovereignty. *Foreign Policy,* pp. 20–29.

Krasner, S.D. 2005. The case for shared sovereignty. *Journal of Democracy*, 16(1), pp. 69–83.

Lamy, S.L. 2011. Contemporary mainstream approaches: neo-realism and neo-liberalism. *IN:* Baylis, J., Smith, S. and Owens, P. (eds.) *The globalization of world politics, an introduction to international relations.* 5th ed. New York: Oxford University Press, pp. 114–129.

Maathai, W. 2010. *The challenge for Africa.* London: Arrow Books.

Martell, L. 2007. The third wave in globalisation theory. *International Studies Review*, 9(2), pp. 173–196.

McGrew, A. 2011. Globalisation and global politics. *IN:* Baylis, J., Smith, S. and Owens, P. (eds.) *The globalization of world politics, an introduction to international relations.* 5th ed. New York: Oxford University Press, pp. 14–33.

McLuhan, M. 1964. *Understanding Media.* Toronto: Signet Books.

McMichael, P. and Schneider, M. 2011. Food security politics and the millennium development goals. *Third World Quarterly*, 32(1), pp. 119–139.

Murphy, S. 2013. *Land grabs and fragile food systems—the role of globalisation.* Institute for Agriculture and Trade Policy.

The Irish Times. 2013. Interview with Pascal Lamy, World Trade Organisation. *The Irish Times: Business,* 19th April 2013, p. 5.

Ohmae, K. 1995. *The end of the nation state: the rise of regional economies.* London: Harper Collins Publishers.

Patel, R. 2007. *Stuffed and starved.* London: Portobello Books.

Rafael, S. and Stokes, D. 2011. Globalising West African oil: US 'Energy security' and the global economy. *International Affairs,* 87(4), pp. 903–921.

Rangnekar, D. 2014. Geneva rhetoric, national reality: the political economy of introducing plant breeders' rights in Kenya. *New Political Economy,* 19(3), pp. 359–383.

Robin, M. 2010. *The world according to Monsanto: pollution, corruption, and the control of our food supply.* New York: The New Press.

Rosenau, J.N. 2000. *The governance of fragmegration: neither a world republic nor a global interstate system* (A Paper Prepared for Presentation at the Congress of the International Political Science Association, Quebec City, August 1–5, 2000). The George Washington University.

Rosenau, J.N. 2003. *Distant proximities: dynamics beyond globalisation.* Princeton: Princeton University Press.

Rosenberg, J. 2000. *The follies of globalisation theory.* London: Verso.

Rosenberg, J. 2005. Globalisation theory: a post-mortem. *International Politics,* 42(1), pp. 2–74.

Sassen, S. 2011. The global street: making the political. *Globalisations,* 8(5), pp. 573–579.

Scholte, J.A. 2008. Defining globalisation. *The World Economy,* 31(11), pp. 1471–1502.

Shiva, V. 2001. *Protect or plunder.* London and New York: Zed Books.

Sklair, L. 2002. *Globalisation, capitalism and its alternatives.* Oxford: Oxford University Press.

Slaughter, A.M. 1997. The real new world order. *Foreign Affairs,* 76(5), pp. 183–197.

Slaughter, A.M. 2004. *A New World Order.* Princeton: Princeton University Press.

Sorensen, G. 2004. *The transformation of the state: beyond the myth of retreat.* Palgrave: Macmillan.

Stevens, T. 2013. *Sovereign data in international relations* [Online]. Available from: http://thesigers.com/analysis/2013/11/11/sovereign-data-in-international-relations.html [Accessed 24th June 2015].

Stiglitz, J. 2005. The overselling of globalisation. *IN* Weinstein, M.M. (ed.), *Globalisation: what's new.* New York: Columbia University Press, pp. 228–262.

Strange, S. 1996. *The retreat of the state.* Cambridge: Cambridge University Press.

Tansey, G. 2011. Whose power to control? Some reflections on seed systems and food security in a changing world. *IDS Bulletin,* 42(4), pp. 111–120.

Tucker, V. ed. 1997. Cultural perspectives on development, *European Journal of Development Research* 8(2), pp. 1–21.

UNCTAD. 2006. *World Investment Report.* Geneva: UN Conference on Trade and Development.

Waltz, K. 1979. *Theory of international politics.* Reading, MA: Addison-Wesley.

Wang, G. 2004. The impact of globalisation on state sovereignty. *Chinese Journal of International Law,* 3(2), pp. 473–484.

Willetts, P. 2011. Transnational actors and international organizations in global politics. *IN:* Baylis, J., Smith, S. and Owens, P. (eds.) *The globalization of world politics, an introduction to international relations.* 5th ed. New York: Oxford University Press, pp. 326–345.

CHAPTER 3

Seed Sovereignty and Globalisation

In this chapter I apply the three different interpretations of the effect of globalisation on sovereignty to the issue of seeds. I divide the chapter into three sections. Section "From Food Security to Seed Sovereignty" begins by distinguishing between the concepts of food security and food sovereignty. It then turns to the distinction between seed security and seed sovereignty. Section "Threats to Seed Sovereignty" identifies various factors influencing change in seed sovereignty. Section "Globalisation and Seed Sovereignty and Security" returns to the interpretations of globalisation and sovereignty presented in Chap. 2 and looks at how hyperglobalist, sceptical and transformationalist schools would understand the exercise of seed sovereignty in the face of these factors.

From Food Security to Seed Sovereignty

Food Security and Food Sovereignty

Interstate food governance was recognised as a 'moral and security imperative' (McKeon 2015, p. 13) by the founders of the Food and Agriculture Organisation (FAO) of the United Nations during a conference of allied governments in Hot Springs, Virginia, in 1943 (Shaw 2007, p. 3). (Figure 3.1 provides a definitional timeline mapping the key events from the emergence of food security as a working concept to the emergence of the concept of food sovereignty.)

© The Author(s) 2019
C. O'Grady Walshe, *Globalisation and Seed Sovereignty in Sub-Saharan Africa*, International Political Economy Series, https://doi.org/10.1007/978-3-030-12870-8_3

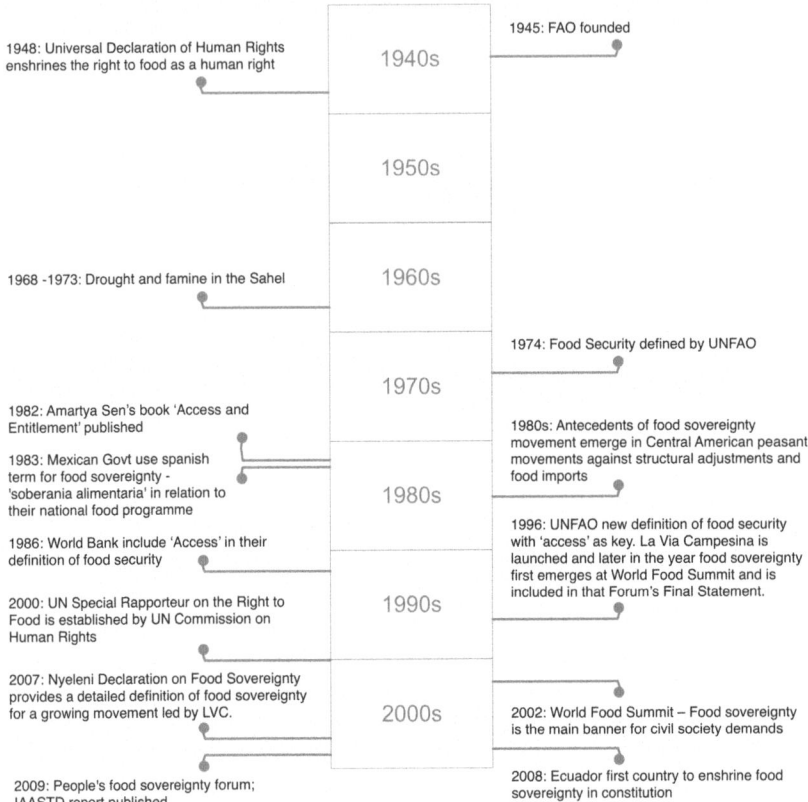

1948: Universal Declaration of Human Rights enshrines the right to food as a human right

1940s

1945: FAO founded

1950s

1968 -1973: Drought and famine in the Sahel

1960s

1970s

1974: Food Security defined by UNFAO

1982: Amartya Sen's book 'Access and Entitlement' published

1983: Mexican Govt use spanish term for food sovereignty - 'soberania alimentaria' in relation to their national food programme

1986: World Bank include 'Access' in their definition of food security

1980s

1980s: Antecedents of food sovereignty movement emerge in Central American peasant movements against structural adjustments and food imports

1996: UNFAO new definition of food security with 'access' as key. La Via Campesina is launched and later in the year food sovereignty first emerges at World Food Summit and is included in that Forum's Final Statement.

2000: UN Special Rapporteur on the Right to Food is established by UN Commission on Human Rights

1990s

2007: Nyeleni Declaration on Food Sovereignty provides a detailed definition of food sovereignty for a growing movement led by LVC.

2000s

2002: World Food Summit – Food sovereignty is the main banner for civil society demands

2009: People's food sovereignty forum; IAASTD report published

2008: Ecuador first country to enshrine food sovereignty in constitution

Fig. 3.1 Definitional timeline: From food security to food sovereignty

When the United Nations General Assembly adopted the Universal Declaration of Human Rights in 1948, the right to food[1] became enshrined in a new post-war framework of basic human rights. This was part of a more general interest in 'human security' (Hettne 2009) that included education, health, welfare and other essential freedoms. Even so, the basic concept of food security was first officially defined by the FAO only in the early 1970s,[2] which speaks to the highly politicised nature of the topic

[1] Article 25 of the Universal Declaration of Human Rights http://www.un.org/en/universal-declaration-human-rights/index.html [accessed online 28 March 2017]
[2] This was at the World Food Conference called by UN in 1974.

(McKeon 2015). At this time, and against a backdrop of Sahelian famine, food security was defined as the:

> availability at all times of adequate world food supplies of basic foodstuffs to sustain a steady expansion of food consumption and to offset fluctuations in production and prices. (United Nations 1975 cited in FAO 2003, p. 28)

This definition was associated with a state-focused programme of price stabilisation and with what Patel (2009, p. 664) calls a "technocratic faith in the ability of states to redistribute resources if the resources could only be made available". In 1976, freedom from hunger became "legally binding when the International Covenant on Economic, Social and Cultural Rights (ICESCR) entered into force" (McKeon 2015, p. 81). According to ICESCR, ratifying states

> recognise the right of everyone to an adequate standard of living for himself and his family, including 'adequate food' and affirming the existence of 'the fundamental right of everyone to be free from hunger'. (ibid.)

In the 1980s, the issue of food security was the subject of further official reports. Following Amartya Sen's (1981) seminal work, *Poverty and Famines: An Essay on Entitlement and Deprivation*, identifying 'access and entitlement' as the primary causal antecedents of poverty and hunger, the Director General of FAO issued a report stating:

> The ultimate objective of world food security should be ensuring that all people at all times have both physical and economic access to the basic food they need ... ensuring production of adequate food supplies; maximising stability in the flow of supplies; and securing access to available supplies on the part of those who need them. (FAO 1983 quoted in FAO 2006, p. 1)

In 1986, the World Bank issued a report titled "Poverty and Hunger"—defining food security in a way that also included 'access', stating: "Access of all people at all times to enough food for an active, healthy life" (World Bank 1986, p. 1). However, this interpretation was restricted by the World Bank/IMF attachment to programmes of structural adjustment in the 1980s and 1990s (Lahiff et al. 2007, p. 1420), with concomitant "deregulation of agricultural markets and dramatic reduction in state support to farmers with differential effects across commodity groups and types of producers" (ibid.). This is seen as part of a "broader development discourse at the time which favoured a greater role for the market and less role for government" (Murphy 2014, p. 226), but Lahiff asserts

> in the main, been highly detrimental for peasant producers and agricultural labourers, contributing to a growing crisis of rural poverty, unemployment and landlessness. (Lahiff 2007, p. 1420)

The Bank themselves admit that this model ultimately "eroded the productive capacity of agriculture" (World Bank 2008, p. 138). By the time the Human Development Report was issued by the United Nations Development Programme (UNDP) in 1994, food security had been relegated back into the broader framework again, and as Acharya (2011, p. 480) points out was now defined within the scope of human security to include seven key areas of security: economic, food, health, environmental, personal, community and political.

Reinvigorated by new leadership at UNFAO, the institution "sought to put food security back on the international agenda", and convened the World Food Summit (WFS)[3] in 1996, which culminated in the adoption of the Rome Declaration on World Food Security,[4] signed by 185 governments and the European Union (EU). A revised definition of food security, with the addition of social access, followed:

> Food security exists when all people, at all times, have physical, social and economic access to sufficient, safe and nutritious food that meets their dietary needs and food preferences for an active and healthy life. (FAO 1996, http://www.fao.org/docrep/003/w3613e/w3613e00.htm)

This was significant as it included 'social' access as well as the nutritional dimension. It is this definition that remains "most broadly accepted" today, according to Clapp (2014, p. 207). In addition, the final WFS document defined the following four pillars in food security: access, adequacy, utilisation and stability (McKeon 2015, p. 75). This definition was later adopted by UNFAO in 2006, reflecting important political victories that affect not just discourse but also practical actions (Murphy 2014, p. 226). It is clear, then, that the concept of food security had been on the agenda

---

[3] WFS was attended by 185 countries and the EU. Signed by 112 Heads or Deputy Heads of Government http://www.fao.org/wfs/index_en.htm [accessed online 1 April 2017].

[4] http://www.fao.org/docrep/003/w3613e/w3613e00.htm [accessed on line 1 April 2017].

for a long time by the mid-1990s, that it had been defined in a number of ways and that it had shaped international action.

The concept of food sovereignty was very different. In 1993 La Via Campesina (LVC) (meaning 'peasant way') was formed. Its inception and subsequent evolution catalysed a global movement for food sovereignty.[5] This movement had its origins in the mid-1980s in the peasant communities of Central America and in their mass mobilisation against harsh structural adjustment programmes and food imports from the USA that were swamping their home markets. Edelman (2014, p. 962)[6] points out that the Spanish term for food sovereignty *soberania alimentaria* predates the English version by a number of years, having been first used in 1983 by the Mexican Government in documents relating to their National Food Programme 'Programa Nacional de Alimentación' (PRONAL) (Edelman 2014, p. 959; Grey and Patel 2015). Regardless of its precise etymological heritage (Patel 2009, pp. 663–665), by November 1996 the concept of food sovereignty, as promoted by LVC, had already made its mark at the international NGO Forum held in parallel with the FAO World Food Summit in Rome. The LVC paper titled "Food Sovereignty: A Future without Hunger" (McKeon 2015, p. 77) saw food sovereignty officially referenced in the Forum's Final statement, making its way into the 'operative action plan' of the World Food Summit Declaration, where the UN High Commissioner for Human Rights was invited "to propose ways to implement and realise these rights" (FAO 1996, para. 61. Objective 7.4[e] http://www.fao.org/docrep/003/w3613e/w3613e00.htm). LVC were challenging the narrow definition of food security, contrasting the two concepts and presenting food sovereignty as the proper evolutionary fulfilment of food security.

> Food is a basic human right. This right can only be realised in a system where food sovereignty is guaranteed. Food sovereignty is the right of each nation to develop its own capacity to produce its basic foods respecting cultural and productive diversity. We have the right to produce our own food in our own territory. Food sovereignty is a precondition to genuine food security. (LVC 1996—http://www.acordinternational.org/silo/files/decfoodsov1996.pdf)

[5] Founded in Mons Belgium in 1993, it represents 200 million farmers from 73 countries on 5 continents with 164 affiliated local and national organisations.
[6] Edelman used Google Ngram Viewer, an online Google search engine that charts word frequency for all publications from 1500 to 2008 to determine the origin of the use of the term 'food sovereignty'.

By 2000 a Special Rapporteur on Food had been appointed within the United Nations and when the next World Food Summit came around in 2002, 'eradicating extreme poverty and hunger' had been established as Goal 1 of the United Nations Millennium Development Goals (UNMDGs) and food sovereignty was the main civil society banner for the Summit. Coordinated through the auspices of an international NGO Food First Information and Action Network (FIAN), which together with LVC and Land Research and Action Network (LRAN) were also vociferously campaigning against the World Bank policy, namely, Market-Led Agrarian Reform (MLAR) at this time (see Lahiff et al. 2007, pp. 1417–1436), used the occasion to strengthen "the legal interpretations of the right to food by extending it beyond simple access to food to include access to productive resources, and beyond individuals to collectivities" (McKeon 2015, p. 82). By this time, therefore, there were two concepts: food security and food sovereignty.

Rosset differentiated the two concepts as follows:

> Food sovereignty goes beyond the concept of food security, which has been stripped of real meaning. Food security means that every child, woman, and man must have the certainty of having enough to eat each day; but the concept says nothing about where that food comes from or how it is produced. (Rosset 2003, p. 1)

Windfuhr and Jansen expanded on this, explaining that "while food security is more of a technical concept and the right to food a legal one, food sovereignty is essentially a political concept" (Windfuhr and Jansen 2005, p. 15).

Raj Patel recognises the intervention of food sovereignty as critical in the definitional debate that was organically emerging around food security, arguing that "critically the definition of food security avoids discussing the social control of the food system" (Patel 2009, p. 665). He goes on to state that under such loose definitions of food security, "It is possible to be food secure in prison or under a dictatorship" (ibid.).

Patel clarifies the precise difference between food security and food sovereignty conceptually—the former being about having enough to eat, pointing out that the conditions surrounding the food getting to the table are not important in this definition, whilst the latter, which is central to my research question, is about who exercises power and control in the food policy space in the face of global forces. While the concepts of food security and food sovereignty can be separated, it is true that some authors contest the 'binary nature' of this distinction (Clapp 2014; Jarosz 2014; Murphy 2014), arguing for the need to 'resist uniformity' (Jarosz 2014, p. 179), to allow for differentiated outcomes in distinct locations. This is a reasonable suggestion in light of the interrelated nature of both concepts and the possibilities which a fluidity surrounding them allow. Nonetheless, in this book I will be specifically looking at the ability to exercise autonomy, power and control in a key aspect of food sovereignty—namely seed sovereignty. It is at this critical point where seed choice and selection is exercised which provides us with an insight into where sovereignty lies.

It is worth noting that the concept of food sovereignty is increasing its political muscle globally. Such was the case when 500 representatives from 80 countries, representing hundreds of organisations gathered in Mali in 2007 and signed the Nyeleni Declaration to "defend and promote the right of people to food sovereignty around the world" and introduced the six pillars of food sovereignty.[7] For the organisations involved, Nyeleni 2007 is regarded as their 'political platform', and a direct challenge to the business-as-usual developmental model of agribusiness.

Ecuador was the first country to enshrine 'food sovereignty' in their national constitution in 2008, with Venezuela, Mali, Bolivia, Nepal and Senegal and most recently Egypt following suit in 2014. By November 2009 the People's Food Sovereignty Forum[8] met in a parallel 'alternative' conference to the UN FAO World Food day ceremony, declaring that food sovereignty:

> Entails transforming the current food system to ensure that those who produce food have equitable access to, and control over, land water, seeds, fisheries and agricultural biodiversity. All people have a right and responsibility to participate in deciding how food is produced and distributed. Governments must respect, protect and fulfil the right to food as the right to adequate, available, accessible, culturally acceptable and nutritious food. (Declaration from Social Movements/NGOs/CSOs Parallel Forum to the World Food Summit on Food Security Rome, 13–17 November 2009, http://www.fao.org/fileadmin/templates/wsfs/Summit/CSO_docs/Final_Declaration-EN.pdf)

[7] http://www.fao.org/3/a-ax736e.pdf
[8] Six hundred representatives of 450 NGOs.

Jarosz's analysis is important here. She sees this rights-based approach to food as an explicit collective vision and objective for the realisation of food sovereignty at transnational, national and local levels, demonstrably directed at a government's duty to fulfil the human right to food and transformation of the current food system (Jarosz 2014, p. 174).

This objective was also finding a voice in more mainstream circles. The publication of the International Assessment of Agricultural Knowledge, Science and Technology for Development (IAASTD) in 2009,[9] which in a press release stated that "business as usual was no longer an option",[10] defined "food sovereignty as preferable to conventional, industrialised agriculture" (Jarosz 2014, p. 174) marked a critical intervention. It was significant not only because of who commissioned it, but because it was the culmination of the work of 400 top scientists from 57 countries, institutional and government policy-makers and multilateral global institutions (ibid., p. 175). The assertion that "diversification of agriculture systems is likely to become an important strategy for enhancing the adaptive capacity of agriculture to climate change" (IAASTD 2009, p. 418) gave food security and food sovereignty equal standing as interrelated concepts, aided by the moral and biophysical realities of climate change.

Food sovereignty is now championed through the reformed and reactivated FAO Committee on World Food Security (CFS), which is the foremost global institutional food forum (McKeon 2015, p. 109). The CFS issued its second draft of the Global Strategic Framework for Food Security in 2012, recognising the "need for consensus on the adoption of the concepts of food sovereignty, with the Nyeleni Declaration's definitions being inserted into the draft" (Jarosz 2014, p. 175). Through an innovative 'civil society mechanism', (McKeon 2015) which allows autonomous civil society involvement in the political process of decision-making, this recognition is unprecedented in UN history (McKeon 2015, p. 108, quoting CFS 2009) and gives disparate civil society and social movements a concrete platform, elevating the standing of groups such as the

[9] The IAASTD is the most recent and comprehensive assessment of agriculture, co-sponsored by the World Bank, FAO, United Nations Environment Programme (UNEP), United Nations Development Programme (UNDP), World Health Organisation (WHO), United Nations Educational, Scientific and Cultural Organisation (UNESCO) and the Global Environment Facility (GEF).

[10] See the brochure summarising the report, p. 4, https://www.globalagriculture.org/fileadmin/files/weltagrarbericht/EnglishBrochure/BrochureIAASTD_en_web_small.pdf, accessed 22 December 2018.

International Planning Committee (IPC) for food sovereignty in representing the needs and demands of LVC and other interested and affected parties throughout the world. Since then, three major colloquia of academics and activists have convened at Yale University (2013), at the Institute of Social Studies in the Hague in 2014 and in the Basque country in 2017[11] to discuss among other things "who is the sovereign in food sovereignty?" and calling for more empirical studies to sharpen the "political and analytical framework of the movement" (McKeon 2015, p. 84).

In conclusion, the concept of food sovereignty has further politicised the discourse around food control and governance. From the grassroots to the highest institutional level at the UN FAO, it has successfully refocused attention back on the central issues of power, control, risks and benefits (Tansey 2011; Scoones and Thompson 2011) in the food political space. In this book I will be specifically looking at the ability to exercise autonomy, power and control in a key aspect of food policy sovereignty—namely seed policy sovereignty, examining where sovereignty lies in decision-making in the seed sector, to which I now turn.

Seed Security and Seed Sovereignty

Seed, described by Kloppenburg (2014, p. 1225) as "the irreducible core of agricultural production", is the basis of our agricultural heritage, dating back 10,000 years to the cultivation of plants at the dawn of civilisation in ancient Mesopotamia. Seeds therefore have "co-evolved with human society and economy as farmers' selections and decisions shaped and continue to shape much of human history" (McCann 2011, p. 24). Unsurprisingly, ensuring an adequate supply of seed has always been closely linked to food security and insecurity. (Figure 3.2 provides a definitional timeline of the origins of the work on seed security to the emergence of seed sovereignty as a working concept.)

Yet, the earliest specific definitions for seed security in its own right at a formal level, only emanate from FAO as recently as 1998, where it is defined as:

> Farming households (men and women) [having] access to adequate quantities of quality seeds and plant materials of adapted varieties at all times—good and bad. (FAO Seed and Plant Genetic Resources Service 1998, p. 187, quoted in McGuire and Sperling 2011, p. 496)

[11] http://elikadura21.eus/publicaciones/

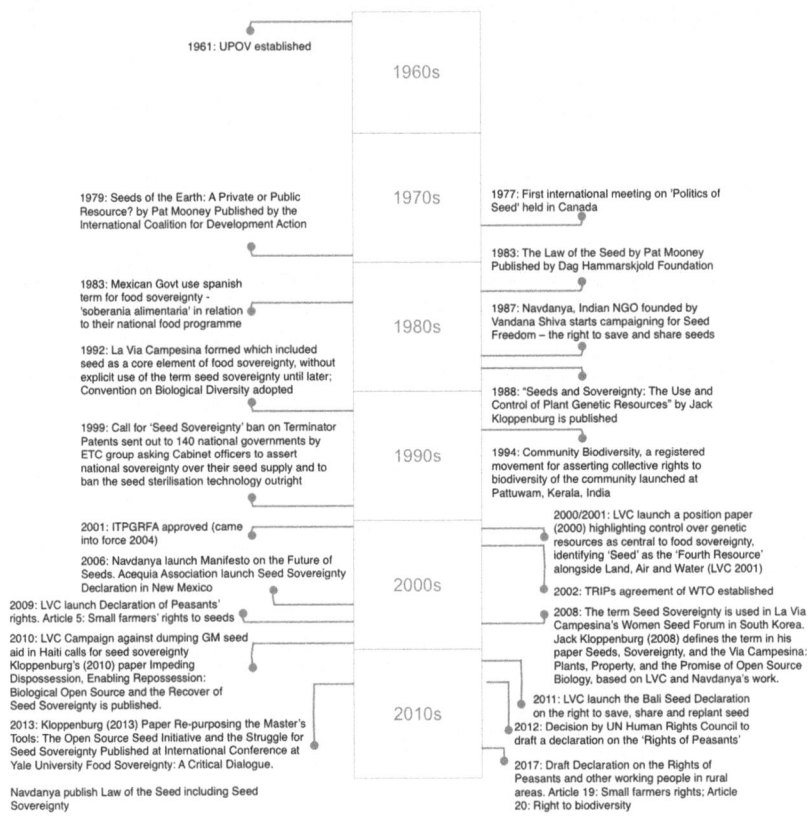

Fig. 3.2 Definitional timeline: Seed security and seed sovereignty

According to FAO (2015), this definition, which is part of the Seed Security Conceptual Framework (SSCF), was inspired by the USAID Food Security Conceptual Framework (1995) (Remington et al. 2002, p. 318), which "provided an improved basis for assessing and analysing seed security and developing appropriate actions to be taken" (FAO 2015, p. 2).

The UN FAO 1998 definition cited above gives substance to the core ideas of availability, access and quality, in ensuring the production and maintenance of seed capacity, seed flows and supply at different levels—household, community, national and regional levels. It is therefore dependent on

Table 3.1 Determinants of seed security

Seed security	Determinants
Availability	Seed close to the farmer (spatial availability) in time for sowing (temporal availability)
Access	Own resources to buy seed or barter for it
Quality	Good quality seed with varietal attributes (like size, etc.) that are acceptable to the farmer

Author's interpretation of Remington et al. (2002, p. 319)

systems of seed 'availability' (from production, trade and transfers), 'access' (entitlements) and 'quality/utilisation' (whether food/seed can meet users' particular 'needs') (McGuire and Sperling 2011, p. 496; Remington et al. 2002). Remington et al. (2002, p. 319). The table below clarifies how seed security manifests itself (see Table 3.1). This shows how availability is determined by both spatial and temporal availability to meet farmers' needs for sowing purposes, by access to cash/finance/vouchers and by quality, whereby the seed is appropriate to the farmers' needs, namely the varietal differentiation is acceptable to fulfil the needs of the household.

Sperling's work and Remington et al.'s framework are instructive. Sperling identifies critical points of difference between seed security and food security, specifically for purposes of disaster management in vulnerable zones. She identifies that "attaining seed security means finding a way to support the systems that give farmers ongoing access to seed of crops and varieties they require" (Sperling 2008, p. 5).

In later work, McGuire and Sperling (2011, p. 497) build on this idea, juxtaposing the food security and seed security frameworks. They highlight the precise points of difference, where availability in food security is simply that food is available close to where people need it, whereas seed security requires spatial and temporal availability for precise sowing requirements. Similarly, seed security definitions of access are slightly more nuanced than food security, as people's ability to produce their own seed is added as well as the capacity to source seed. Utilisation is even more nuanced in the differentiation between food security and seed security here, as there is greater emphasis on farmers' preferences for specific seed varieties, whose quality best suits the requirements of the farmer.

McGuire and Sperling argue for special assessments to be conducted to ensure that "food security assessments should no longer be used as proxies (implicitly or explicitly) for seed security assessments" (ibid., p. 502).

The differences identified clearly instructed UN FAO's more recent definition of seed security, which describes it as

> ready access by rural households, particularly farmers and farming communities, to adequate quantities of quality seed and planting materials of crop varieties, adapted to their agroecological conditions and socioeconomic needs, at planting time, under normal and abnormal weather conditions. (FAO 2014)[12]

This definitional adjustment is significant, because, as McGuire and Sperling (2011) identified, "food need does not translate to seed need" (McGuire and Sperling 2011, p. 502), as even in situations of chronic stress, seed may be available when food is not. More importantly, this definition gives expression to the ability to access seed that is adapted to distinct agroecologies (a term associated with the food sovereignty discourse). It also recognises that the 'informal' seed system is also an inherent part of social systems, especially in the Global South, where access and entitlement to seed is largely determined by family and community ties, differentiated socio-economic networks, through barter systems, gift exchanges and/or community markets.

Thus, while there is a difference between seed security and food security generally, both concepts, as understood here, are related to the idea of freedom from want, and whilst the most recent definition of seed security gives an indication of how or where those seeds may be sourced, it does not spell out entirely how that freedom might be achieved.

The term 'seed sovereignty' has only entered the academic lexicon in recent years, predominantly through the work of Jack Kloppenburg in his detailed unravelling of the constituent parts of the definition as it has emerged mainly within agrarian movements like LVC (2008). Kloppenburg (2008, p. 3) points out that seed sovereignty had resided in the hands of farmers and gardeners until the 1930s, when the advent of more profound agricultural changes began a process of dislocation of seed sovereignty which I shall return to in the next section. In the present period, the escalation of genetic erosion and seed diversity led to efforts to reassert sovereign control over seed. To this end, Kloppenburg formulated the core principles of seed sovereignty in his (2008) paper titled "Seeds,

---

[12] http://www.fao.org/agriculture/crops/core-themes/theme/seeds-pgr/seed_sys/security/en/

Sovereignty, and the Via Campesina: Plants, Property, and the Promise of Open Source Biology" (2008). This was based on the work of the two main organisations working in the seed space, namely the Indian-based seed NGO, Navdanya (meaning 'Nine Seeds'), which had been founded by scientist Vandana Shiva in 1987 to protect and enhance indigenous seed and crop varieties from corporate capture in agriculture and for seed freedom, and LVC,[13] who since 1992 had been advocating for food sovereignty, where seed sovereignty is recognised as the 'fourth resource', alongside land, air and water (LVC 2001). Though the term 'seed sovereignty' was not explicitly used by these organisations in their earlier published work, it nevertheless was implicit and indeed central to their vision and mission, and it was from their materials that he extracted the fundamental principles and features and formulated what he calls a "set of linked features which together constitute a coherent and robust structure" (Kloppenburg 2010, p. 385). This allowed him to use the term as a "heuristic domain" to place the commonality of perspectives from both organisations (Kloppenburg 2014, p. 1234), and in doing so established a more coherent frame of reference for what we can say constitutes seed sovereignty.

That said, the battle for what we now understand as seed sovereignty had been erupting for decades, albeit under different titles such as plant genetic resource (PGR) conservation, farmers' rights, access and benefit sharing and agrobiodiversity conservation. Canadian agricultural expert, Pat Mooney (1979, 1983), is regarded across the literature as a pioneer in this area, as he examined the issue of plant genetic resource conservation in his two published works, *Seeds of the Earth: A Private or a Public Resource (1979)* and *Law of the Seed (1983)*, and convened the first international conference on the 'politics of seed' in 1977 in Canada under the auspices of Rural Advancement Foundation International (RAFI)[14] now called Erosion, Technology and Concentration Group (ETC). This coincided with the already alarming rate of loss of genetic diversity, as the Green Revolution's commercial seed paradigm took hold. Mooney and a central group of key seed activists, scientists and others within and outside

---

[13] https://viacampesina.org/en/index.php/actions-and-events-mainmenu-26/stop-transnational-corporations-mainmenu-76/904-haitian-peasants-march-against-monsanto-company-for-food-and-seed-sovereignty [accessed online 19 March 2017].

[14] RAFI (now ETC) was the first CSO nationally or internationally to draw attention to the socio-economic and scientific issues related to the conservation and use of plant genetic resources, intellectual property and biotechnology.

the UN FAO in particular, acted as the catalyst to much of the work on seed sovereignty which followed and were instrumental in the high-level global battles for control over seed which have since been erupting at various international fora. Esquinas-Alcazar (2005, p. 949) highlights that as early as 1967 and subsequently in 1973 and 1981, the FAO had been establishing 'technical advances in relation to PGR conservation', culminating in FAO establishment of the intergovernmental Commission on Genetic Resources for Food and Agriculture (CGRFA) in 1983[15] and leading to a 1989 'agreed interpretation' of FAO International Undertaking on Plant Genetic Resources, and in 2001 the approval of a highly contested International Treaty on Plant Genetic Resources for Food and Agriculture (ITPGRFA) (Kloppenburg 2010, p. 373), which came into force on 29 June 2004 (Esquinas-Alcazar 2005, p. 950). This Treaty, despite its critics,[16] established a Multilateral System of Access and Benefit Sharing, through which "sovereign nations have agreed to share resources and benefits" (ibid.) related to a wide body of seed/plant varieties and also made provision for farmers' rights.

This work was also central to the adoption by UN Environment Programme of the Convention on Biological Diversity (CBD) in 1992, "the first binding international agreement of its kind" (Esquinas-Alcazar 2005, p. 949), signed by 140 parties, whose main aim was "ensuring the conservation of biological diversity, the fair and equitable sharing of benefits arising out of the use of PGRs" (Article 1), and the maintenance of in-situ and ex-situ conservation measures (Articles 6, 7, 8) (De Schutter 2009, p. 7), all foundational features of seed sovereignty. However the "Treaty left most of the implementation regarding smallholder farmers' rights to seeds to the discretion of national governments" (Coordination Sud 2017, p. 12).

Simultaneously, other negotiations were afoot. Seed had become a transworld mobile technological 'artefact' with the Agreement on Agriculture (AoA) in the Uruguay round in the mid-1990s. A heretofore unmoveable agricultural sector was 'deterritorialised' (Scholte 2008) and opened to the vagaries of a global market, followed by various globalising laws such as the intellectual property regimes such as the TRIPS Agreement of WTO in 2002 and the "Union Internationale pour la protection des

---

[15] Esquinas-Alcazar (2005) states that "CGRFA provided the first permanent international forum for the negotiation, development and monitoring of international agreements and regulations in this field" (p. 949).

[16] The criticism relates to its abuse by TNCs.

obtentions végétales" (UPOV 91). UPOV, which originated in France in 1961, was conceived and designed by European commercial breeding interests and instituted Intellectual Property Rights for plants and plant breeders (Berne Declaration[17] 2014; Dutfield 2011; Tansey 2011). This effectively "institutionalised the process of agricultural liberalisation on a global scale by restricting the rights of sovereign states to regulate food and agriculture" (Holt Gimenez and Shattuck 2011, p. 111). It was at this time that we encounter the first direct call for seed sovereignty through the successful campaign in 1999 by ETC,[18] led by Pat Mooney, calling on all governments to ban Terminator Patents (seed sterilisation technology) in letters sent to 140 governments asking Cabinet officers to "assert their national sovereignty over their seed supply and to ban the seed sterilisation".[19]

In recent years the farmers' rights provisions, which are inextricably linked with seed sovereignty, have been revisited. UN Special Rapporteur for food, Olivier de Shutter (2009), sharpened the focus on the direct relationship between seed policies and food security especially in vulnerable communities in the Global South. He pointedly raised concerns fundamental to the seed sovereignty debate, including the "risk of Intellectual Property (IP) related monopoly rights neglecting poor farmers' needs, undermining traditional systems of seed saving and exchange, and losing biodiversity" to the "uniformisation encouraged by the spread of commercial varieties" (De Shutter 2009). This intervention coincided with the publication of the IAASTD report in 2009, which specifically addressed the essential 'conservation through use' of plant genetic resources and was a significant moment for the cause of seed sovereignty. This also coincided with LVC's 2009 adoption of a Declaration on Peasants' rights. After seven years of consultation, Article 5 stipulated 'small farmers' rights to seeds. By September 2012, this led directly to the decision by UN Human

---

[17] Berne Declaration, now called Public Eye, is a Swiss-based organisation focussing on business and human rights.

[18] ETC (Erosion, technology and concentration) Group, originally known as Rural Advancement Foundation International (RAFI) is an international civil society organisation (CSO) that addresses the global socio-economic and ecological issues surrounding new technologies with special concern for their impact on indigenous peoples, rural communities and biodiversity. It was the first CSO to draw attention to the socio-economic and scientific issues related to the conservation and use of PGRs, IP and biotechnology.

[19] http://www.etcgroup.org/fr/content/call-seed-sovereignty-ban-terminator-patents [accessed online 20 April 2017].

Rights Council to draft a Declaration on the 'Rights of Peasants'. The latest draft (officially since adopted in November 2018) contains Article 19 'small farmer right to seeds' and Article 20 the right to biodiversity (Coordination Sud 2017, p. 14).[20]

By 2013, Vandana Shiva more explicitly addresses the term 'seed sovereignty': "Seed for the farmer is not merely a source of future plants/food; it is the storage place of culture, of history". She defines seed sovereignty as "the right of farmers to save, use, exchange, and sell their own seeds" (Shiva 2013). This echoes closely the definitions put forward by LVC:

> LVC see the *practice of seed sovereignty, through open, free, non-commoditised exchange as not only si*gnifying the centrality of seed in the production of food, but in the very reproduction of culture itself. (Carolan 2012, pp. 262–265)

At this point, many other groups were also actively working on the issue of seed sovereignty and conservation of the integrity of genetic resources, including ETC, GRAIN, Bioversity International, Greenpeace as well as many grassroots organisations and community seed-saving organisations and seed banks emerging around the world.

This all fed in to a later publication by Kloppenburg titled "Re-purposing the Master's Tools: The Open Source Seed Initiative and the Struggle for Seed Sovereignty" (2014), in which he condenses the four principal and constitutive elements of what he terms 'seed sovereignty' based on their work, as follows:

1. The right to save and replant seed
2. The right to share seed
3. The right to use seed to breed new varieties
4. The right to participate in shaping policies for seed (Kloppenburg 2014, pp. 1234–1235).

He goes on to identify four key areas, or what he calls 'foundational principles of seed sovereignty'. These are:

1. Community seed saving and exchange—'in-situ'—dynamic conservation of farmer cultivars,

---

[20] This has been voted on twice where interestingly EU has shifted from voting against (June 2014) to abstaining (October 2015) (C2A Notes, Coordination Sud, No. 232015).

2. Agroecology and participatory plant breeding (PPB),
3. Legal sovereignty over seed—a concrete juridical mandate,
4. Openness to allies (ibid., p. 1236).

Seed sovereignty, as clarified by Kloppenburg here, is therefore not about how many are being fed or the nutrition levels of the food people are eating, though these are all important issues in their own right. It is how those people are making choices in relation to the kind of seed/food they are sowing, reaping and eating—the degree of autonomy and access they have in that process, and why they are making the choices they do. This is what Da Via calls the "articulated autonomy of local communities", expressed through their ability to "access a diversity of products and methods of production and innovation, as well as to a diversity of purposes and venues of exchange" (Da Via 2012, p. 238).

But the extent to which communities or indeed nations can control those seed choices has significantly increased the stakes in the battle for seed sovereignty in the face of globalisation. For Kloppenburg:

> Seed is the critical nexus where contemporary battles over the technical, social and environmental conditions of production and consumption converge and are made manifest. Who controls the seed gains a substantial control over the shape of the entire food system. (Kloppenburg 2010, p. 368)

McKeon highlights that now under pressure from farmers' organisations and other civil society participants, the Governing Body of the International Seed Treaty (ITPGRFA) meeting in September 2013

> adopted a resolution renewing governments' commitment to implement the 'Farmers' Rights' provisions introduced into the 1986 Treaty negotiations thanks to determined civil society advocacy. (McKeon 2015, p. 125)

The resolution specifically seeks to defend farmers' rights to save, use, exchange and sell farm-saved seed; to be recognised and rewarded for their contribution to the global pool of genetic resources; and to participate in decision-making on issues related to crop genetic resources (Mulvany in McKeon 2015, p. 125). Increasingly, authors and a wide variety of organisations, including (ACB 2017, 2015; Oakland Institute 2017; AFSA and GRAIN 2015; Munyi 2015; Berne Declaration 2014; World

Bank 2013a[21]; IAASTD 2009) have voiced concerns regarding the risk attached to jeopardising farmers' seed systems, all in one way or another highlighting core principles of seed sovereignty, even if in the case of the World Bank, for example, they are also clearly considered to be part of the problem of undermining those same principles (Oakland Institute 2017; AFSA and GRAIN 2015).

In conclusion, I have mapped out the trajectory of concerns surrounding food security to the emergence of a specific push for food sovereignty and similarly tracing the seed security debate leading to more recent discussion surrounding seed sovereignty. This book is about seed sovereignty. It examines how global forces are affecting actors' ability to act independently in their seed choice and use, with particular reference to those vulnerable regions, such as subsistence farmers in Ethiopia and Kenya.

THREATS TO SEED SOVEREIGNTY

This section identifies the major threats to the exercise of seed sovereignty at the global, national and local level. These threats are all interrelated and the multidisciplinary nature of the scholarship in the area affirms the cross-sectoral context of the subject and the importance it holds globally. Different authors address these threats in different ways, and with different emphases. I summarise this vast literature by identifying seven major threats to seed sovereignty.

Threat 1: Climate Change

Climate change is characterised by erratic weather patterns, with dramatic fluctuations in temperature, and rainwater variability, leading to extremes of drought, storms and flooding worldwide (IPCC 2014[22]; FAO 2011).[23] Climate change has the potential to affect seed sovereignty both directly and indirectly.

Firstly, "catastrophic extreme weather events can pose immediate threats to the survival of breeds and varieties that are raised in specific

---

[21] http://siteresources.worldbank.org/EXTNWDR2013/Resources/8258024-1352909193861/8936935-1356011448215/8986901-1380046989056/WDR-2014_Complete_Report.pdf

[22] http://ipcc.ch/pdf/assessment-report/ar5/syr/AR5_SYR_FINAL_SPM.pdf [accessed online 30 March 2017].

[23] http://www.fao.org/3/a-i6030e.pdf [accessed online 30 March 2017].

small geographical areas" (FAO 2016, p. 28).[24] Specialised monocultures and ex-situ developed seed, which are characteristic of 'modern' mass production agricultural systems, are particularly susceptible due to their narrow genetic variability and reduced capacity for adaptation to changing climatic conditions (FAO 2011; International Assessment of Agricultural Knowledge, Science and Technology for Development [IAASTD] 2009).[25] This represents a critical threat to seed sovereignty, as it may force some people out of farming altogether, due to the sheer impossibility of growing anything, especially in the most volatile weather zones exposed to maximum stress such as sub-Saharan Africa. This is a direct loss to sovereign seed systems and the continuity of seed/farming knowledge, what Carolan refers to as biocultural diversity (Carolan 2012, p. 160).

Secondly, vulnerable farmers are also more susceptible to 'novel' seed interventions particularly from global corporations tied to government and some aid sectors who are promoting 'climate-smart seed inputs'.[26] These 'climate-ready crops' (ETC 2010) are being offered as a solution to climate change. ETC point out that the world's six largest agrochemical and seed corporations are presently "pressurising governments to facilitate the broadest and potentially most dangerous patent claims in intellectual property history" (ibid., p. 1). Three companies—Du Pont, BASF and Monsanto—account for 66% of these patent claims (ibid.), marking a concentration of corporate power which will further undermine the rights of farmers to save and exchange seeds. Thus, both directly and indirectly climate change can have an effect on seed sovereignty.

[24] A 20% decrease in growing periods is projected for parts of sub-Saharan Africa. Sixty-five countries in the 'South' risk losing 280 million tonnes of potential cereal production, valued at $56 billion as a direct result of climate change (FAO 2005 quoted in ETC 2009).

[25] The IAASTD is the most recent and comprehensive assessment of agriculture, co-sponsored by the World Bank, FAO, United Nations Environment Programme (UNEP), United Nations Development Programme (UNDP), World Health Organisation (WHO), United Nations Educational, Scientific and Cultural Organisation (UNESCO) and the Global Environment Facility (GEF).

[26] The term 'Climate smart agriculture' was first used by UN FAO in 2010 as a means to attract corporate finance for agriculture in Africa. See [https://www.grain.org/article/entries/5270-the-exxons-of-agriculture] accessed online 31 March 2017.

Threat 2: Conflict and Displacement

Wars and conflict pose a problem for seed sovereignty in two ways.

Firstly, conflict can cause the displacement and/or death of farmers, and scientists, whose lineage of knowledge is critical to maintaining seed practices. It can destroy land and seed banks, either through direct chemical warfare and landmines or through the destruction of critical locations of seed accessions.

Conflict directly causes a loss of knowledge, the disappearance of bio-cultural memory, culinary traditions and a lineage of seed stewardship—that is, seed sovereignty (Carolan 2012, p. 160; Nabhan 1997, p. 2). Ethno-botanist, Gary Nabhan states:

> Where human populations had stayed in the same place for the greatest duration, fewer plants and animals have become endangered species; ... where massive in-migrations and exoduses were taking place, more had become endangered. (Nabhan 1997, p. 2)

This is borne out by accounts from various conflicts and disaster-related migrations such as Rwanda, Ethiopia, Iraq and Afghanistan. In the case of Iraq, the war resulted in 600,000 farmers being unable to farm, whether due to landmines, enforced migration or actual war, making the threat of loss of seed knowledge and seed stewardship directly observable.

Secondly, seed sovereignty can be affected by post-conflict developments. For example, conflict has led to the imposition of new seed laws and policies, bringing fundamental changes in seed and agricultural practices and posing a serious threat to seed sovereignty. For example, a new Seed Patent law called 'Order No. 81' was introduced in Iraq in 2004[27] and 'Seed Law 2006'[28] in Afghanistan. Both laws were passed during periods of conflict and military occupation and were not agreements made between sovereign nations. Both seed laws had considerable impact on the exercise of seed sovereignty by local populations and experienced critical changes in seed production, use and control (Hassan 2005; Focus on the Global South and GRAIN 2004). For example, Order 81 was part of a

---

[27] "Patent, Industrial Design, Undisclosed Information, Integrated Circuits and Plant Variety Law of 2004, CPA Order No. 81, 26 April 2004", http://www.iraqcolaition.org/regulations/20040426_CPAORD_81_Patents_Law.pdf [accessed online October 2010].

[28] http://www.wipo.int/edocs/lexdocs/laws/en/af/af006en.pdf [accessed online 31 March 2017].

suite of laws (100 in total) with the ultimate intention of "privatisation to promote economic diversity"[29] (US State Department documents online).[30] The new law rendered it illegal for Iraqi farmers to reuse seeds harvested from new varieties registered under the imposed law. It also paved the way for genetically modified crops or organisms into Iraqi agriculture for the first time.

Threat 3: Changes in Farming Practices

Changes in agricultural practice pose a problem for seed sovereignty in at least three ways.

Firstly, 'labour-saving devices' have caused both an expansion of farm size and a concomitant shrinking of the 'commodity basket' (Carolan 2012, p. 16). This has resulted in a dramatic diminishing of on-farm seed diversity, access and knowledge. It also led to a separation and compartmentalisation of aspects of agriculture into units of specialisation, which increasingly could be dealt with off-site or controlled in new mechanised systems.

Secondly, seed sovereignty was seriously threatened with the advent of the 'Green Revolution' in the late 1960s, backed by a neo-liberal economic paradigm. The Green Revolution is best described as a

> series of research and technology transfer initiatives, funded by large US private foundations like Ford and Rockefeller, centred primarily on the development of high-yield varieties of a handful of grains. (ibid., p. 73)

Coupled with the expansion of the necessary irrigation infrastructure and 'input supply chains' (fertiliser, pesticides, seeds), this 'modernising paradigm' in farming practice both transformed agriculture, increasing productivity and yields, whilst simultaneously contributing to a significant decline in seed sovereignty. For example, the Philippines grew over 3000 varieties of rice prior to the introduction of the Green Revolution in the 1960s. Twenty years later, there were only two rice varieties on 98% of the total land area, with a worldwide erosion of crop diversity of 75% for the twentieth century (FAO 2010; Berne Declaration 2013).

---

[29] https://www.globalpolicy.org/war-and-occupation-in-iraq/37145.html [accessed online 31 March 2017].

[30] http://govinfo.library.unt.edu/cpa-iraq/index.html [accessed online 31 March 2017].

Thirdly, this mechanisation and widespread erosion of seed sovereignty in the Western world has brought a corresponding decline in the number of farmers (Patel 2007, p. 40). In the 1930s 25% of the US population lived on six million farms. Today two million farms are home to just 2% of Americans, with 8% of farms accounting for 72% of sales (ibid.). The seismic shift away from agriculture and farmer seed has brought huge social change, but also a loss of local knowledge and practices.

Threat 4: Recent Technological Changes

Scientific discoveries pose a problem for seed sovereignty in two ways.

Firstly, developments in plant genetics in the twentieth century brought new highly bred homogeneous seed varieties (F1 hybrids),[31] generally referred to as improved varieties, to fulfil the need for uniformity, productivity and the growing market for monoculture cash crops. They have underpinned the formal seed sector for commercial agriculture for the last half century in the Western world and are associated with increased yield and productivity in intensive agricultural systems, which are dependent on irrigation and agrochemicals. In more industrialised countries they have displaced the Open-Pollinating Varieties (OPVs), which are locally adapted heterogeneous 'landrace'[32] varieties of seeds that when mature can produce seed that can be saved again (the system of 'in-situ' farming and gardening that has been the hallmark of agronomic systems since wild plants were first cultivated in the Fertile Crescent 10,000 years ago). However, F1 hybrids posed a serious challenge to seed sovereignty. A key difference between OPVs and F1 hybrid seeds is that F1 hybrids are not as resilient in the second generation and so are not appropriate for seed saving; thus, the farmer is forced to return to the commercial seed owner to purchase the next season's stock. Therefore, the F1 hybrid seeds, despite any short- to medium-term benefits they may bring, lock farmers into contractual

---

[31] F1 hybrid refers to filial generation 1. The first filial generation of offspring of distinctly different parental types.

[32] A landrace is a variety of domesticated animal or agricultural plant species which has developed over a long period of time and as a result has adapted to the local natural environment in which it lives. Landraces are selected and grown from seed passed down from generation to generation and thus exhibit broad heterogeneity associated with wider genetic diversity and therefore greater adaptability and resilience, now considered critical to maintaining the genetic resource base for food security (Altieri 2009).

seed arrangements and represent a step away from total seed sovereignty, as the lineage of stewardship is broken.

Secondly, the development and commercialisation of Genetically Modified Organisms (GMOs) or transgenic seeds and Biotechnology (BT) in the late twentieth century marked another critical juncture within agriculture, with far-reaching consequences for seed sovereignty. This scientific and technological shift marked a profound change involving the intentional manipulation of genes from different species using biotechnology to insert particular traits from one species into another, creating a transgenic plant. The small number of very powerful corporations, mainly Monsanto, which have been commercialising these seeds in agriculture since 1996 claim that these seeds have the potential to substantially increase the yields and variety of foods available globally, as well as support the ambitious agricultural export plans for poorer countries (Robin 2010; ETC 2009; Patel 2007). GM seeds represent the antithesis of seed sovereignty, as these seeds are laboratory-produced, corporate-owned, with strict prohibitions on resowing. Olivier De Schutter, as UN special rapporteur on Food, outlined how the advent of these broader patenting and IP rules accompanying the introduction of biotechnology and GM has totally altered the farming relationship and threatens to entirely undermine the principles of seed sovereignty:

> Farmers cultivating patented seeds do not have any rights over the seeds they plant. They are considered to be licensees of a patented product, and they frequently are requested to sign agreements not to save, resow or exchange the seeds which they buy from patent-holders. (De Schutter 2009, p. 6)[33]

These new proprietorial technologies look set to continue to be a key threat to seed sovereignty globally.

Threat 5: TNCs

TNCs can be defined as corporations which are registered and operate in more than one country at a time. They can pose a problem for seed sovereignty in two ways.

[33] UN General Assembly A/64/170 Item 71 (b) The Right to Food [Accessed online 23 February 2013].

Firstly, the increasing power and influence of a few giant seed corporations due to consolidation, monopolisation and concentration of their market power greatly skews the emphasis on profits over protection of practices related to seed sovereignty. Market rights prevail over farmers' rights as TNC dominance in the global seed market continues apace. It has increased in value to $37 billion in recent years (FAO 2009),[34] with cross-border seed trade worth $6.4 billion at 2007 levels. The global market for maize seed marketed as 'drought tolerant' is an estimated $2.7 billion, while the US Department of Agriculture predicts that the global bio-based market for chemicals and plastics alone will top $500 billion per year by 2025 (USDA Foreign Agricultural Service 2008).

Allied to this is the increased TNC consolidation through mergers and acquisitions, particularly since 1996 which has led to increased monopolisation of power within seed/chemical TNCs. In 1996 the ten biggest seed companies had a market share of less than 30% of the global proprietary seed market. Today, "six firms, all based in the West, currently control over two-thirds of the formal seed-market" (ETC 2015, cited in the Oakland Institute 2017, p. 3). In the case of Monsanto, a recent report by the Berne Declaration states that:

> Through the acquisition of Seminis, the world's biggest producer of vegetable seeds for US$1.4 billion in 2005, Monsanto became the global market leader for vegetable seeds as well. (Berne Declaration 2013, p. 10)

In the grain sector this consolidation is even greater. Three giant US TNCs, Cargill,[35] Archer Daniel Midland (ADM) and Bunge, control 90% of the world's grain (Magdoff and Tokar 2010). This TNC consolidation is leading to soaring profits but is accompanied by a significant negative correlation for both biodiversity and seed sovereignty for key crop varieties in many locations. Organisations such as ETC, GRAIN, Association for Plant Breeding for the Benefit of Society (APBREBES) and Oakland Institute regularly publish materials monitoring the pace of mergers and acquisitions in the food industry, highlighting that it rose to $4.5 trillion in 2007, doubling nearly every two years since 2000 (ETC Group cited in Magdoff and Tokar 2010, p. 20), and is marked by consolidation and control horizontally and vertically well beyond just the agribusiness sector, and well beyond farming. It extends now from chemicals,

---

[34] Second World Seed conference Rome 2009.
[35] Cargill's revenue for 2012 was $133.9 billion.

fertilisers, seeds, to processing, retailing, packaging and exporting, all dominated increasingly by a diminishing number of corporations as monopolisation of control from seed to supermarket continues unabated (Magdoff and Tokar 2010). The most recent proposed merger between Monsanto and Bayer:

> if approved the new merged company would control almost 30% of the global commercial seed market and 25% of the agrochemical market making it the world's largest supplier of seeds and chemicals. (African Centre for biodiversity, ACB 2017, p. 4)

This is set in the context of other recent mergers between China National Chemical Corporation (ChemChina)—Syngenta and DuPont and Dow. Therefore if this merger goes ahead, "just three corporations would control about 60% of the global patented seed market and 64% of the agrochemical market" (ibid.).

Secondly, seed sovereignty is directly threatened by GMOs, which is dominated by one TNC, Monsanto. Monsanto seeds now occupy 87% of the total area devoted to GMOs worldwide (New Internationalist 2010, p. 10), thereby emerging as the biggest player, followed by Du Pont, Syngenta, Dow and Bayer. The top three are also the leaders in the pesticide market further consolidating their power over the value chain, global agricultural systems and seed sovereignty (Berne Declaration 2013, p. 2, p. 9).

Threat 6: Legal Threats to Seed Sovereignty

Legal changes pose a problem for seed sovereignty in two major ways.

Firstly through launching seed into the speculative markets which occurred with the passage of the Agreement on Agriculture (AoA), one of the Uruguay Round Agreements and a central plank of the trade agreements which established the WTO in 1995. The AoA marked a profound shift in policy as agriculture had always been isolated from wider trade negotiations. Murphy (2013, p. 5) points out that because "land cannot be moved" and even food because of issues of durability was restricted from wider trading, the mobility central to all aspects of the shrinking world of globalisation was restricted in the food space. But seed has mobility and is much more susceptible to intervention. The AoA helped to open up the seed space to economic forces, launching the food arena into the speculative markets of private capital and the dynamics of the capitalist system, with a plethora of new rules attached (Murphy 2013; Tansey 2011). It effectively "institutionalised the process of agricultural liberalisation on a global scale by restricting the rights of sovereign states to regulate food and agriculture" (Holt Gimenez and Shattuck 2011, p. 111). Seed sovereignty was now directly challenged by this global agreement.

Secondly, changes in laws surrounding seeds and plants, specifically around patents and proprietorial control spread from the USA through the Organisation for Economic Co-operation and Development (OECD) countries, and now acts as a global driver in the agriculture and food sector, with profound implications for seed sovereignty, especially in the area of biotechnology and genetic engineering, as

> agrichemical companies with their patent based chemical business background were drawn into the messy longwinded art and science of plant breeding. (Tansey 2011, p. 113)

The TRIPS Agreement, which was enacted in 2002, was central to shifting proprietorial legal control because it made it "illegal for farmers to reuse seed without permission or payment" (Tansey 2011, p. 114). Article 27.3 (b) of TRIPS requires WTO Members to "provide for the protection of plant varieties either by patents or by an effective *sui generis* system or by any combination thereof" (quoted in Dutfield 2011, pp. 10–11). TRIPS also brought the International Convention for the Protection of New Varieties of Plants known as (UPOV 91) centre stage. It was already an established legal instrument since 1961, with an in-built system of plant variety protection (PVP), a form of intellectual property for plant varieties. UPOV already had 67 member states signed up, with all the major commercial powers, except India. So when UPOV intervened in the WTO Council for TRIPS in 2002 stating that UPOV Convention met the requirements of 27.3 (b) of the TRIPS Agreement, it gave the impression that UPOV membership was essential for TRIPS compliance, which, though false as many authors point out, pressurised poorer vulnerable countries in the Global South and other states to introduce UPOV in order to harmonise with obligations for WTO accession (Berne Declaration 2014; Dutfield 2011; Downes 2003).

The UPOV system of PVP and plant-breeders' rights (PBRs) together with TRIPS represents the single biggest shift in power and control in seed sovereignty away from community/public ownership to commercial private ownership. UPOV insists on the 'DUS' (distinct, uniform and stable) standard for plants. Because of this requirement, protection is not extended to farmers' varieties (generally landrace varieties and Open-Pollinated Varieties [OPVs]), which though considered to be "inherently unstable and in permanent evolution" (De Schutter 2009, p. 6), are central to seed sovereignty and are integral to seed and food security for agrarian communities throughout the Global South in particular. These rules became an institutionalised legal driver creating an enclosure by private capital within the seed space (Dutfield 2011; Tansey 2011), including the "privatisation and specialisation of agricultural research" (Da Via 2012, p. 230), heightening tension between proponents of IP on the one hand and vast numbers of small farmers and breeders as well as advocates for conservation of biodiversity and seed sovereignty on the other (Berne Declaration 2014; De Jonge 2014; Downes 2003).

Threat 7: Tied Aid and Philanthrocapitalism

Philanthrocapitalism can be defined as a 'hands-on venture philanthropy' which directly and substantially affects public policy unlike previous philanthropic endeavours, such as the famous Carnegie Foundation libraries of the previous century. It is "distinguished by the goal of remaking the public sphere in its own image" (Thompson 2014, p. 5). It seeks to:

> mirror the business world, with a quantifiable results based tabulation system, focussing on capacity building in shorter multi-year programmes, rapid response system in a climate where their financial aid is so desperately needed. (Kumashiro 2012, p. 15)

Philanthrocapitalism poses a problem for seed sovereignty in two main ways.

Firstly, it is seen as a specific new threat to the exercise of seed sovereignty, as it represents a significant shift in autonomy away from the public realm, the 'commons', into the hands of private actors, effectively removing autonomy from a key public realm and placing public seed policy into private hands.

Many authors refer to the new and powerful influence of foundations like Alliance for a Green Revolution in Africa (AGRA), a partnership between the Rockefeller Foundation and Bill and Melinda Gates Foundation (BMGF) since 2006. AGRA describes itself as an independent organisation, based in Africa and led by Africans, whose primary strategy is to facilitate the creation of an efficient African food system through grants and capacity-building assistance to institutions that are helping to improve the productivity of smallholder farmers. (Gates Foundation and AGRA 2015[36])

They have a special interest in increasing farmer's access to 'good seeds', including "developing entrepreneurship in marketing high quality hybrid seeds" (ibid.). As one of the 'anchor grantees on the African continent', having expanded its donor base to include governments and international organisations (ibid.), its growing power in what was once a publically controlled seed space is significant.

Secondly, philanthrocapitalist/new private actors have considerable vested interest in commercial aspects of the seed arena themselves, yet are not accountable to the affected public/communities in sovereign states where they act. These philanthrocapitalists are engaged in profit-making ventures with corporate seed companies themselves, yet are playing a role in dismantling farmers' seeds systems and public seed bodies.[37] This has the combined threat of removing the participatory nature of plant breeding from a key social realm of farmer practice, but also removing systems of accountability which heretofore would have rested within public/state/parliamentary systems of governance (Thompson 2014, p. 5).

In conclusion, it is evident that seed sovereignty has faced and continues to face numerous challenges. Much of the erosion of seed sovereignty has occurred in the pursuit of a model of development and modernisation in agriculture, which for much of human history has gone uncontested. This is no longer the case, as many of these threats now highlight a growing cross-disciplinary concern about the loss of crop genetic diversity, and concomitant shifting of seed sovereignty from the public/commons to private corporate interests. In the next section I present how the various schools of thought look at how actors are affected by globalisation and how they address the changes in seed politics, seed policy and seed sovereignty.

[36] www.gatesfoundation.org [accessed online 19 March 2017].
[37] Examples are BMGF 2010 multi-million dollar purchase of Monsanto shares and Rockefeller Foundation's vested interest in its own 'novel' seed varieties being registered in Kenya, with vast royalties in the pipeline.

Globalisation and Seed Sovereignty and Security

In Chap. 2, I distinguished between the hyperglobalist, sceptical and transformationalist perspectives on globalisation and sovereignty. In this chapter I apply these perspectives to the domain of seed sovereignty. As before, I focus on three key areas, namely, the state, the role of IOs and the role of TNAs. Here, I present how writers understand how globalisation affects sovereignty in the seed space in these three areas. There is a relatively small literature that directly addresses seed sovereignty and none which evaluates the relationship between seed sovereignty and global forces using this model, despite indirect references and analysis. So, using the heuristic device of the three schools I am extrapolating from across a wide literature in order to evaluate how seed sovereignty is faring. Table 3.2 summarises the positions.

Table 3.2   Three perspectives on globalisation and seed sovereignty

	State	*IOs*	*TNAs*
Hyperglobalists	Deterritorialisation—Transboundary nature of seed mobility Erosion of state/public role in seed policy/programme	UPOV, WTO, TRIPS, AoA set rules backed by WB IMF. Harmonisation of laws	Core duties and functions derogated to others—especially TNCs Global penetration mergers and acquisitions Corporate control—Monsanto, Du Pont, Gates, AGRA and others
Sceptics	State is still the main driver of policy choices and policy implementation	IOs do what big states tell them	TNAs still look to strong 'Northern' states that determine outcomes
Transformationalists	'Multiple equilibria' State is a disaggregated player but still has a role in certain aspects Glocalisation—local initiatives	Ambiguities within World Bank and FAO re IAASTD report Contradictions within UN role and substance CBD, IPCC	Multidimensional not unidirectional Technological innovation can go either way Homogeneity and heterogeneity

Hyperglobalists

As we saw in the previous chapter, hyperglobalists hold two distinctly different normative perspectives, namely globophobia and globophilia (Held and McGrew 2007, p. 2). The globophobic authors claim that globalisation, particularly market globalism, is sweeping the state aside and that this is a negative development. Globophilia represents the neo-liberal capitalist modernisation perspective, which sees globalisation as a principally democratising and emancipatory force, through its dominant drivers—'technics— the combined force of IT and the communications revolution' (Held and McGrew 2007). Both agree that a profoundly 'new' phase of global intensification and integration is sweeping the old status quo aside, most notably state sovereignty. It is clear that the normative element is more profound in the hyperglobalist literature than in either the sceptical or the transformationalist schools. This is borne out when assessing the hyperglobalist literature in relation to seed sovereignty. Here, I examine what hyperglobalist authors say about seed sovereignty in the context of the domestic state, IOs and TNAs. It is worth noting, as before, that some authors articulate the hyperglobalist position in their critique, but overlap into other schools in their prescription for change.

The Domestic State

Many globophobic authors now identify the agricultural sphere in general and the politics of seed in particular as one of the most important theatres where the battle for control over globalisation is occurring (George 2015; AFSA and GRAIN 2015; Shiva 2005; McMichael and Schneider 2011; ETC 2010; Patel 2009). They identify the erosion of the state/public role in seed policies and seed programmes as a key signal of the state being released of its 'core functions' (Strange 1996). Murphy (2013) identifies a key moment when agriculture became a transworld, marketable commodity, when seeds, 'the irreducible core of agricultural production', became a mobile artefact through key institutional changes in the 1990s, namely the AoA, a core element of the establishment of WTO as detailed earlier. This represented a classic manifestation of everything hyperglobalists recognise as the essence of globalisation's disregard for state power. The deterritorialisation of seed opened the way for global corporate appropriation, a new colonialism, a deliberate dispossession of states and peoples of their national collective seed/food sovereignty (Kloppenburg 2014). The new IP rules system is seen as facilitating "a private system of taxation—as firms controlling key patents, copyright and trademarks can set prices as they see fit to benefit their own interests" (Tansey 2011, p. 115), sweeping the state aside at every turn, as evidenced by the relentless pursuit of 'seed harmonisation laws' and policies transworld now (AFSA and GRAIN 2015, ACB 2015; Oakland Institute 2017; Tansey 2011). These hyperglobalists see the domestic state as increasingly irrelevant, as 'shadow sovereigns' take their place (George 2015).

C.B. Thompson, a globophobe in my terms, identifies philanthrocapitalism, as the newest manifestation of this neo-liberal food regime, which she calls an "assault on government participation in any economic sphere" and sees it as "global and comprehensive since 1980" (Thompson 2014, p. 6). She states:

> 'It began with removing government from the production sector, declaring all parastatals as inefficient ... incapable of organising production relations'. ... In agriculture it meant the removal of any state farms, from USSR to Mozambique and Zimbabwe, even ones experimenting with new varieties of seeds or rationalising production on degraded soils. (ibid.)

Thompson highlights the deep roots of this new cosmocracy in places like Southern Africa, extending as it did to

> removing government partnerships in seed breeding, a long tradition of public sector interest for national food security. ... Credit, agricultural research and extension all fell victim to the neo-liberal project of reduction of government intervention. (Thompson 2014, pp. 6–7)

Globophiles on the other hand argue that globalisation accelerates agricultural production, stimulates markets and will lift people out of poverty (Langyintuo 2011; Lamy 2011). It can erode the state and government, which act as unnecessary "middlemen" to global solutions (Ohmae 1995, p. 4). Globophilists in the seed space regularly refer to the stifling 'bottlenecks' in the system of government, consider the state and parastatal agencies as overly bureaucratic and inefficient, hindering "the growth of the seed sector and consequently depriving farmers of the benefits of genetic improvements" (Langyintuo 2011, p. 6).

Globophilist hyperglobalists are behind the push for corporations, and/or philanthrocapitalists, to take over where states once led—seeking "better coordination and regulations across countries to speed up the introduction of much needed technology for use by African farmers" (Syngenta quoted in *The Guardian* 2014). Interventions, especially the promotion of public-private partnerships (PPPs) in the seed space, are seen as essential in redefining agricultural development agendas, with reconfigured roles for producers, the private sector and the state (Scoones and Thompson 2011, p. 5). This shift is considered by globophiles to be essential for global governance of global problems such as hunger, food security, biodiversity and climate change in particular, but essentially, and controversially, serves the needs and wishes of global market capitalism.

Both globophobes and globophiles believe that new actors are assuming authority over a once key function of state power, the ability to control seed choice and selection to sustain food production and security within state borders for local populations, whilst also shifting the terms of their trade in food and seed with other states.

International Organisations

The development of key international organisations marks a distinct new period which sees authority being superimposed by international organisations over the state in financial control, technological 'developments' and law, all now exercising globalising control over seed sovereignty (AFSA and GRAIN 2015; McKeon 2015; McMichael and Schneider 2011; Harrison 2010).

Globophobic hyperglobalists see this domination by a few key IOs, and the force of globalising seed laws, including IPRs, PVP, PBRs through UPOV, AoA and later TRIPS as instrumental in consolidating power and control of global seed systems within a new globalising corporate regime (George 2015; Kloppenburg 2014; Murphy 2013; Shiva 2005; Dutfield 2011; McMichael and Schneider 2011; Tansey 2011). Many commentators identify key supranational financial institutions, specifically the IMF and World Bank, with their high-impact structural readjustment policies as central to the private sector and particularly detrimental to agriculturally dependent poorer countries. These IOs, they say, exposed the Global South to unfair competition with subsidised products from abroad. This increased their indebtedness, causing further severe cutbacks of state support to and regulation of agriculture. This in turn has set the stage of what has come to pass—states reduced to the service of markets, governed by supranational institutions who set the rules (McKeon 2015, McMichael and Schneider 2011; Maathai 2010). The globophobes hold the FAO and World Bank in particular responsible for backing the replacement of 'low-yielding' traditional varieties (of seed) with 'high-yielding' varieties developed by international agricultural research systems alongside their national counterparts in the early days of the Green Revolution. This led to the dismantling of the public breeding programmes, followed in the 1980s and 1990s with the privatisation of the state seed companies, and new laws and regulations sweeping away "trade barriers, thereby encouraging or forcing farmers to buy certified seed every year" (GRAIN 2005, p. 28).

This was facilitated by the "World Bank's financing of policy changes and projects on the ground ... targeting land and seed laws as key tools for protecting the interests of the corporate sector" (AFSA and GRAIN 2015, p. 5). It was backed by other key institutions, such as International Finance Corporation (IFC), International Rice Research Institute (IRRI), Consultative Group on International Agricultural Research (CGIAR), European Bank of Reconstruction and Development (EBRD), Challenge Programme of Harvest Plus, the International Centre for Agricultural Research in the Dry Areas (ICARDA), the International Crops Research Institute for Semi-Arid Tropics (ICRISAT) and Comprehensive Africa Agriculture Development Programme (CAADP), all of which "favour the private interests of the agricultural industry" (Thompson 2014, p. 10). Key foundations, which I shall return to in the next section, AGRA and the Gates Foundation, are also considered to be central players in this new era of 'unequal comparative advantage' with agriculture shifting from the public sphere into private hands, particularly in sub-Saharan Africa (McMichael and Schneider 2011, p. 123).

However, the establishment of the WTO in 1995 is considered as a key 'moment' when an IO, namely the WTO, criminalised seed saving and sharing. This was seen as a direct attack on seed sovereignty, causing a deep disaggregation of core state functions, as WTO affects much of what occurs in the seed space from the formal state level down to local household/farm and market decisions (Holt Gimenez and Shattuck 2011). For Shiva, the institutions of the Uruguay Round, General Agreement on Tariffs and Trade (GATT), WTO, TRIPS and the AoA are the IOs which she has labelled 'biopiracy' (Downes 2003, Chap. 4). Shiva describes this as:

> the corporate privatisation of a genetic strain, without recognition or 'benefit sharing' to those who cultivated it over millennia, and criminalised actions to protect the biological and cultural diversity on which diverse food systems are based. (Shiva 1993 quoted in Downes 2003, p. 32)

This was seen as an outright rejection of the concept of 'reciprocity', enshrined in the international legal principle of Access and Benefit Sharing, first put forward by another important global agreement, the Convention on Biological Diversity (CBD) in 1993, Article 8 (Thompson 2014, p. 3). Patel points out that the aim of the EU and USA during these negotiations was "to keep their own strategic reserves of food, while forcing countries in the Global South to cede sovereignty over their agricultural supplies" (Patel 2007, p. 97).

For globophobes, this IO takeover of the seed space is inherently dangerous, because of its association with speculation in commodities which exacerbated the 2007–2008 food crisis, an ongoing crisis regarded as 'historical and systemic' (McMichael and Schneider 2011, p. 120). It also pitches the international agricultural research centres of CGIAR against the breeding programmes of National Agricultural Research Institutes (NARIs), which, they assert, are causing considerable decline in public sector capacities in many countries as a result (Scoones and Thompson 2011, p. 5).

Thompson cites the case of ICRISAT in Zimbabwe which in 2010 "ceased sharing foundation seed with the smallholder farming sector, from whom the germplasm came in the first place", as well as the violation by ICARDA of the non-exclusion rights in the case of beer barley, when they allowed an exclusive private-sector ownership and control of ICARDA's advanced barley lines (i.e. seeds) to three major brewers, one owned by Heineken and two in which Anheuser-Busch has 50% ownership (ETC 2010, p. 20). Thompson sees this as the corporate sector trying to transform a specific international law of 'no exclusive rights', inherent in CGIAR rules, into "its own image of exclusion, for a short time to encourage innovation … seeking to transform free sharing of seeds into privatisation, in a new way, beyond patenting" (Thompson 2014, p. 11).

Globophiles see these developments as positive for trade and investment. It furthers global integration, breaking down barriers to commerce with the radicalising potential to encourage human freedom and prosperity without the stifling public bureaucracy and the power politics of states (Held and McGrew 2007, p. 189). Globalisation of the seed space was a positive development in their eyes, as it gave mobility to a commodity which had heretofore been geographically fixed, within state structures and agricultural mindsets, which to them were disabling and inefficient. The new globalising drive articulated by the World Bank Group Action Plan 2013–2015 gives emphasis to "improving the resilience of agricultural systems and rural livelihoods through support for more 'climate smart' agriculture". They specifically call for International Bank for Reconstruction and Development/International Development Association/International Finance Corporation (IBRD/ IDA/IFC) agricultural lending and investments that support climate change adaptation and mitigation, such as the development and adoption of more drought and flood-tolerant plant varieties (World Bank 2013b, p. xvii), all of which speak to the globophile position of institutionalised, corporatized, commoditised seed system agendas taking over, where states once led.

For hyperglobalists, the borderless market has reached the seed. This is witnessed through the transworld imposition of a new rule system orchestrated by a supranational institutional framework which has incorporated the seed system into its orbit of influence, a profound disaster for humanity for globophobes and an enormous success of a global project for globophiles. Through the actions and policies of key IOs and their globalising legislative appendages—IP, PVP, PBRs—hyperglobalists point to the increased regulatory and phytosanitary restrictions on seed on the one hand to fulfil commercial seed requirements of uniformity and distinction for the global food industry, and deregulation of markets and trade barriers on the other, all as evidence that global forces have assumed sovereignty over the seed.

Transnational Actors

Many hyperglobalist authors recognise the increased emphasis on "partnerships between governmental, paragovernmental and non-governmental organisations", with no preferential position for the state (Jessop 1997, pp. 574–575). This is clearly borne out in the seed space for many authors, highlighting a growing transworld porosity and deterritorialisation with the intensification of globalisation (Kloppenburg 2014; Murphy 2013; Scholte 2008). They identify a wide range of new transnational actors who are assuming roles and responsibilities in seed law, policy and practice at every level (selection, production, distribution, use, propagation, ownership), once held in stewardship by farmers and in public systems run by state, regional and local administrations, whilst also identifying increasingly global responses of resistance to such changes in the transnationalisation of civil society movements.

For globophobes there are three main groupings of transnational actors. Firstly, there are TNCs. These are the key powerful agrichemical

TNCs in the seed space, such as Monsanto, Du Pont, ADM, Bayer, Syngenta and Cargill, some of whom constituted the intellectual property committee. Hyperglobalists point to this lobby's successful inclusion of TRIPS in the establishment of the WTO in the 1990s (Downes 2003), which they say opened the door to creating a 'seed enclosure' by them with unprecedented market power, and access to the genetically diverse 'Global South'. They have since consolidated their power, with mergers and acquisitions in the seed industry, profits from which are used to build the next level of intensification and consolidation for their industry (ACB 2017; Oakland Institute 2017; McMichael and Schneider 2011, ETC 2010). Armed with globalised legal instruments, particularly IPRs, the seed space is regarded as "driven by companies rather than countries" consistent with globophobia (Held and McGrew 2007, p. 39). This is evident in increasing monopolising control of the entire value chain from farm to fork (AFSA and GRAIN 2015; McKeon 2015, Berne Declaration 2014; ETC 2010), and with specifically tightening control of seed use and distribution in the formal sector, which is now at the heart of negotiations on global trade rules and central to membership and accession to the WTO.

Secondly, they identify the huge private philanthrocapitalist foundations, such as the Rockefeller Foundation and BMGF, as a 'new' phase, bringing new actors, seeds/products/inputs and with unprecedented power and agency at the deepest level of penetration yet, right to the seed, homed in the cultural and spiritual lifeworld of the remotest communities (McCann 2011; Shiva 2005; Odame and Muange 2011; Alemu 2011; Maathai 2010). McKeon highlights how BMGF:

> made its first grant in the field of agriculture in 2006, coinciding with their alliance with the Rockefeller Foundation forming AGRA, whose objective was to deliver the Green Revolution in Africa. By the end of 2009, Gates had invested more than $1.4 billion in promoting a 'new green revolution strategy' whose implementation is accountable solely to its two co-chairs. This compares—in terms of democracy and transparency—with FAO's regular budget of $1billion for the 2010–2011 biennium which is debated, approved, and monitored by its 192 member governments. (McKeon 2015, p. 25)

This is considered as evidence that states have been devoured by new private transnational actors.

Similarly, McMichael and Schneider state:

AGRA's role can be distinguished by this privatisation of modernisation, deepening the application of agri-technologies (including biotechnology), reorienting agriculture as an export industry producing non-traditional crops (e.g. shrimp and soybeans instead of pineapples and coffee in the Global South for world markets. (McMichael and Schneider 2011, p. 122)

Consistent with hyperglobalist thinking they see it as a new strand, a distinct new layer of intensification, different from before, saying:

It is basically a second Green Revolution but different from the first which involved States in economic nationalist models of modernisation, character-ised by import protection on staple commodities (ensuring food security), and parastatals providing various forms of research, credit, transport and processing assistance. (ibid.)

For McCann it is even more absolute. It represents, he states:

what may well be a sea change in the genetic makeup of African cereal seeds. Smallholder farmers in Africa will be consumers of seed as a finished indus-trial product, not selectors of phenotypic traits they favour locally. African farmers will plant crops with traits isolated directly via DNA extractions in corporate genetic labs located in Iowa, Switzerland, Germany or South Africa rather than by breeders or farmers working in African fields … wrap-ping African farmers in networks of finance, markets and technology transfer that are regional and global and assume the permeability of national borders and ecologies. (McCann 2011, p. 34)

These issues were central to the establishment of the Convention on Biological Diversity (Article 1) 1992, whose aim was to protect domestic seed sovereignty and farmers' rights.

The third group of TNAs are constituted by a plethora of INGOs in the food/seed sovereignty network. These include organisations such as LVC, Navdanya, African Biodiversity Network (ABN), Alliance for Food Sovereignty in Africa (AFSA), the Green Belt Movement, ETC, GRAIN, Oakland Institute and Greenpeace, which are increasingly global in focus and operating outside of state boundaries and structures, and which enjoy a potentially global audience, particularly through global media networks. They operate through new forms of legitimacy with a self-appointed mandate for change in the politics of seed, with the potential to under-mine and even dismantle the agendas of other major TNAs and IOs, despite having less financial resources.

For hyperglobalists, these three groupings are the key transnational seed actors now, driving and determining significant changes. All the while, globophiles regard these changes as progressive. US agri-giant Cargill's Chairman bluntly calls this drive "the commercialisation of pho-tosynthesis" (cited at https://www.cobrt.com/archived-blog/2011/07/trusting-photosynthesis, accessed 22 December 2018). The Cargill logic, which epitomises the globophilic position, is 'trade', "not growing food for local consumption, but growing what you grow best and trade it as widely as possible to your global consumer base" (McMichael and Schneider 2011, p. 127). Former WTO chief, Pascal Lamy, points out that the "developing countries GDP surpassed the developed world in 2012 for the first time in centuries" (*The Irish Times* 19 April 2013, p. 5), a recognition that free trade of goods and services are drivers of prosperity. Others point out that estimates of between 8% and 11% growth rates over 5 years for countries like Ethiopia in sub-Saharan Africa and a reduction in poverty and food insecurity (UNDP-UNDAF 2011) point to huge posi-tives in the 'development model' being pursued. Ethiopia, they say, is a recognised "top mover" worldwide in UNDP Human Development Index (HDI), its growth recognised as the highest in Africa by the IMF (USAID 2012). For globophile organisations such as AGRA or philan-throcapitalists like Bill Gates, it is good business, good science and stan-dardised global technological solutions to global problems—the golden era of globalisation's potentiality being realised right down to the smallest seed in the remotest village.

Sceptics

Sceptics do not dispute the facts of globalisation, but question the claim that global forces are eroding state power to exercise sovereignty, saying that 'globalisation is oversold' (Stiglitz 2005, p. 229). The state remains perhaps not the sole agent, but certainly the main agent (Hay and Marsh 2000) in determining how power is exercised and resources (in this case seeds), are distributed. The bottom line for sceptics is that 'States still mat-ter' (Held and McGrew 2007), acting as 'egoistic rationalisers', promot-ing their own self-interest, be they a major power, a power bloc or a rent-seeking state bartering aspects of sovereignty for comparative advan-tage. For sceptics, the power politics of seed sovereignty and security is no different. International organisations and other actors still look to the state for legitimacy, support and funding. There is a wide-ranging theoretical and normative base to the sceptical school, but a dearth of sceptical literature on seed sovereignty, so I adapt sceptical scholarship to the seed literature here. I organise the following three sub-sections as in the previous section, looking at the domestic state, IOs and TNAs.

The Domestic State

Sceptics say that certain states' interests lie behind the changing seed political landscape as key states and the institutions and organisations they control act as the central organising force in a new world order determining seed security. Food has long been seen as 'a tool' by many political sceptics and practitioners, so it stands to reason that control over vital biological resources, such as seeds, would be considered an important determinant for future vested interests of those states, because, as Gilpin says, "nation-states want the absolute benefits of a global economy at the same time that they seek to increase their own relative gains and protect themselves against external threats to their economic welfare and national security" (Gilpin 2001, p. 357). In the tradition of Hobbes, Carr, Morgenthau and Machiavelli, this is "conflict of interest based on power not harmony of interest based on morality, which defines international order" (Korab-Karpowicz 2010). Seed politics and control of plant genetic resources is increasingly a key feature of powerful states' geopolitical manoeuvring.

Historical materialists, like Harvey on one end of the sceptical spectrum, and Haas and Krasner on the other, agree that the state is the central organiser and enforcer. Harvey's thesis that the core dynamic of the current dominant orthodoxy of neo-liberalism, based on 'accumulation by dispossession', is realised "as in the past, through the power of the state which is frequently used to force such processes even against popular will" (Harvey 2005, p. 161). Harvey specifically includes intellectual property, genetic materials and "seed plasmas" as part of this "commodification, corporatisation and privatisation of nature", all of which are realised or brought to fruition through the now liberalised state, which becomes the "prime agent of redistributive policies" (ibid., p. 163). Contrary to the hyperglobalist claim of seed exemplifying the globalist agenda of deterritorialisation, "territoriality of jurisdiction is still central, despite global forces", according to Harvey, particularly when talking about rights, such as human rights, right to food and the rights associated with dispossession of resources and genetic materials. Since the state retains "its monopoly over legitimate forms of violence, it can in Hobbesian fashion define its own bundle of rights and be only loosely bound by international conventions" but ultimately the state is the one body with the power to enforce (ibid., pp. 178, 180). Harrison, whom I identify as a sceptic, points out:

> A great deal of FDI that has entered Africa in the last twenty years has developed on the back of strong pressures by Western States (and IGOs) to privatise public utilities and state-owned enterprises and to set up international tenders for transfer of ownership. (Harrison 2010, p. 14)

Many authors point out that differentiated investment streams particularly evident with the rise of Brazil, Russia, India, China and South Africa (BRICS), especially China, and their investment in seed markets, emphasise the centrality of the state in shaping international investment deal brokering (Future Agricultures Policy Brief 2014). This, sceptics say, is reinvigorating state power, especially in certain African countries, as emerging powerful states and changing geopolitics provide more options to national/recipient governments in negotiating terms of agreements, resource extraction and aid, all central to the changing seed politics in the region (Future Agricultures 2014; Carmody 2011; Alden 2007).

Sceptics point to the distinct national programmes and applications of seed laws and policy. These range from Chavez's assertion of food sovereignty and seed protection legislation in Venezuela (Article 305, 1999) following his election, while recent seed legislation in Ethiopia underlines the assertion of 'tight state control' against private actors (Alemu 2011) in more robust protection of indigenous seed practices (Proclamation 782/2013, Part 1, Sec 3). In Malawi, local elites are said to ease the passage of corporate interests in adopting hybrid maize seed and subsidy programmes (Chinsinga 2011), while the Ghanaian 'state has been transformed into a regulator of seeds and a catalyst for the emergence of seed markets' (Amanor 2011, p. 48). Of critical importance to the sceptical school is the number of countries that are enshrining food sovereignty in their constitutions in recent years. Beginning with Ecuador in 2008, Venezuela, Mali, Bolivia, Nepal, Senegal and most recently Egypt in 2014, with more countries lining up, all reinforcing the sceptical position of the central role the state still plays in grounding global forces in distinct national settings.

International Organisations

The central argument of the sceptical school is that globalisation is merely a fig leaf to cover up the strategic goals of dominant states, predominantly the USA, and that IOs are the mechanism which they use to consolidate their hegemony.

The institutionalisation of free trade and structural adjustment dominated the post-World War II period. The 'Washington Consensus' is the understanding that the IOs were not only situated geographically in Washington, but also that the IMF, World Bank and UN were politically and ideologically influenced by their country of residence (McKeon 2015; McMichael and Schneider 2011; Harrison 2010). McKeon, whom I identify as a sceptic, points out that the UN FAO was stripped of any real teeth early in its history where "food security impinged on trade" and what were seen to be 'national interests' of the USA and the UK (McKeon 2015, p. 13) and the World Food Board was totally undermined. "Britain and America were not prepared to give either funds or authority to an organisation of which they did not have full control" (McKeon 2015, p. 14, quoting Boyd Orr). By 1971 McKeon points out that:

> the establishment of the Consultative Group on International Agricultural Research (CGIAR) had already excised agricultural science from the UN family and headquartered it at the World Bank. (McKeon 2015, p. 16)

The establishment of UPOV in 1968, and subsequent amendments in 1972, 1978 and 1991 complete with a new regime of plant variety protection and plant-breeders' rights, which favoured OECD, EU and US dominance in agricultural science and law, is seen as critical to many sceptical interpretations of the power play at work in the seed space (Dutfield 2011; Tansey 2011; Robin 2010). Its extension into the heart of WTO negotiations with member countries since the 1990s is seen by sceptics as the 'vested interests' of dominant states to spread influence and maintain their hegemony.

The perspectives vary widely between those who observe how powerful states operate in the international arena, via the international institutions over which they maintain substantial leverage, and those on the other end of the sceptical spectrum who see the politics of seed through the lens of historical materialism—the logic of power and capital being asserted over the rights of citizens and nature (Harvey 2005). Intellectual Property Rights, central to the IOs in the seed space (UPOV, TRIPS, WTO), are but an extension. Echoing Shiva, Harvey refers to the WTO's act of 'biopiracy', when seed became 'private property' in the 1990s (Harvey 2005, p. 160), legitimised since in the policies and programmes of the leading

IOs and dominated by Western, particularly US interests. There is nothing new in that for sceptics. It is simply a continuation of earlier versions of 'realpolitik'.

GRAIN (2005) point out that intervention in the seed space, particularly in Africa, provides an illustrative example of how powerful states (predominantly the USA) operate, setting policy agendas, using the major international organisations as legitimation tools where strategy necessitates. Since the early 1990s USAID has introduced "reforms" (Title 111 Program 1992) under the Clinton administration, followed by "The Harmonisation of Seed Policies and Regulations in the Eastern Africa Project", which was piloted in East Africa in 1999, and is coordinated by ASARECA (Association for Strengthening Agricultural Research in Eastern and Central Africa), the regional umbrella of the national agricultural research services. This strategy of 'harmonisation' is funded by USAID, and part of the World Bank's SSASI (Sub-Saharan African Seed Initiative) project (GRAIN 2005, p. 29). The objective of ASARECA, alongside their policy wing, East and Central Africa Programme for Agricultural Policy Analysis (ECAPAPA), is to harmonise and rationalise seed laws and policies in the region (Minde 2004, p. 1). The pilot programme in Kenya, Uganda and Tanzania subsequently expanded to 'second-tier countries' in 2002, and third-tier countries in 2003, reorienting seed policy to fit the agenda of a dominant state (USA) and a key IO (World Bank) with responsibility for agricultural policy, practice and funding.

In short, dismantling domestic seed networks, through the leading IOs and international law, through structural readjustments, economic liberalisation or outright war, as in Iraq, creates opportunities for the strong state actors that lie behind the forces of change in the seed space to create a new opportunity for a new framework, which best serves their interests, and strategically ties in with other economic, military and foreign policy agendas also. This is consistent with sceptical thinking that these organisations and the powerful states behind them orchestrate, manage and control crises, both to rationalise the system and to redistribute assets (Harvey 2005, p. 162). However, sceptics also point to variation in state adaptation and application which signifies state choice/sovereignty in the seed space, despite the seemingly dominating forces of hegemonic powers backed by key IOs. Its application on the ground varies hugely, as evidenced in the preceding section, suggesting state filtering of new globalising seed laws to suit individual state needs.

Transnational Actors

The sceptical school does not deny the power and force of the transnational food and seed corporations nor the plethora of NGOs, philanthro-capitalists and other transnational actors which globalisation in this sphere has brought in its wake. They simply regard them as offshoots of the prevailing logic of power and capital enshrined in the state structure. For sceptics, these transnational actors are in effect reinforcing the dominance of states and are manifested in two key areas: (1) in TNCs, which are just Trojan horses for states; and (2) in the existence of state-level variation in the exercise of seed sovereignty.

Sceptics claim that powerful states are the decision-makers behind the mega-corporations which are borne out by many authors' assertion of the revolving door between industry, banks and governments which perpetuates dominant structures of power, all the time playing advantage to strong states. McKeon states that "what governance of markets exists, is in the hands of select groups like the G8, G20, and the OECD" (McKeon 2015, p. 206). I already noted that the intellectual property committee of just 13 (all US) MNCs ensured that TRIPS was included on the WTO agenda (Downes 2003). This is significant in light of sceptics' claims that TNCs are still geographically grounded in countries and few if any are truly transnational (Sklair 2002), but rather bat for the big powers in a symbiotic relationship that maintains systems of power and control. This dovetail between certain TNCs, such as Monsanto, Du Pont and Cargill in the seed space and key powerful states such as the USA remains a significant part of the complex interaction between the emerging science of plant genetics, TNCs and the state.

For Harvey, the "pillaging of the world's stockpile of genetic resources for the benefit of a few large pharmaceutical companies" as seeds became private property within the TRIPS Agreement and the WTO was nothing new (Harvey 2005, p. 160). It came in the tradition of state power, which sees class, property rights and capital accumulation as primary. Intellectual Property Rights and privatisation of the commons were just an extension of that state power.

Sceptics would point to TNC support for big government initiatives or their role in promoting legislative and policy seed changes in different countries, particularly the Global South as testament to their position. A clear example would be the case GRAIN highlight when in 1999, the American Seed Trade Association (ASTA) set up the African Seed Trade

Association (AFSTA) as a local lobby for the transnational seed industry (GRAIN 2005). AFSTA was mandated to

> "promote regional integration and harmonisation of seed policies and regulations supportive of US seed trade", with an explicit target of securing a 5% increase in US seed exports to the region in the first 5 years, a direct policy agenda to bolster (US) state hegemony. AFSTA and its 18 national seed industry associations are now deeply involved in all of the major regional and seed law processes. (GRAIN 2005, p. 31)

In a more recent report, GRAIN highlights that:

> since the late 1990s, the US and Europe have been pushing bilateral free trade agreements (FTAs) into Africa as tools to gain market advantages for their TNCs, which directly affects seeds. (AFSA and GRAIN 2015, p. 15)

Similarly, there are many examples of Monsanto's considerable financial support for the G8 New Alliance for Food Security and Nutrition, the Grow Africa Partnership and the New Vision for Agriculture Initiative.

Yet the varying application of new seed laws and policies in different countries reinforces the sceptical argument that 'states still matter' and that most transnational actors dance to a large extent to the tune of the state, be it the dominant hegemon or the more impoverished state. This variation is important. In some countries, the privatisation of the seed industry has been relatively simple, as private investment has moved to replace the state and run viable large private seed companies (Amanor 2011, p. 48). For example, AGRA provided technical and financial support to the Ghanaian Government to "review its seed policies", resulting in new pro-business seed laws in 2010 (Plants and Fertiliser Act 2010) (Act 803) (AFSA and GRAIN 2015, p. 16), while Alemu highlights that "Tensions exist between the Ethiopian state and the emergent private sector as the state seeks to liberalise the sector, while retaining a strong hold over the market" (Alemu 2011, p. 76). For sceptics these variations in state adaptations and applications of seed law and policy reinforce the primacy of the state as the key actor, even if the state acts as 'egoistic value maximisers' (Lamy 2011), bargaining aspects of their seed sovereignty to confer advantage elsewhere.

The Transformationalists

Lastly I return to the transformationalist school of thought. I identify those authors who argue from this position in the seed space. They see the world as essentially globalising, with the food and seed sector as open as everything else to the new borderless world. They also see the adaptive power of the state, and its ability to adjust and reform structures to include new dimensions of governance, central to realism. It is the multidimensional nature of these shifts in the seed space that defines their transformational character, given that "globalisation is a contingent historical process, replete with conflicts and tensions" (Held and McGrew 2007, p. 213), with an inherently uncertain roadmap. I examine what transformationalist scholarship is saying about the impact of globalising seed agendas on each of the three sub-sections, the domestic state, IOs and TNAs. There is a strong normative dimension to most transformationalist scholarship. Whilst acknowledging this, I will lay out the key features which define transformationalism.

The Domestic State

Transformationalists point to evidence of 'glocalisation' (Cerny 2010; Hettne 2009) in the seed space, distinct locales where local and global forces collide, where actions are precisely defined by this 'impact on the local society' (Hettne 2009, p. 87), with the concomitant understanding that it is not unidirectional, and that the local society is itself then simultaneously inserted into globalisation. For Hettne, this period marks the 'second great transformation', in deference to Polyani's *Great Transformation*. He says that specifically "from a cultural perspective globalisation is more complex, giving birth to hybrid forms" (Hettne 2009, p. 88). Transformationalists would point to spaces where in some cases rich agrobiodiverse farming is continuing and even escalating, with new, in some cases state-backed participatory community seed banks emerging, while other places sometimes within the same region or jurisdiction are experiencing almost total erosion of seed diversity and practices, with monocropping and homogeneous agricultural systems increasingly controlled by external forces, and state agencies being established and financed by philanthrocapitalists (Odame and Muange 2011; Scoones and Thompson 2011; Tansey 2011). Transformationalists point to the multinucleated, multilevel nature of globalisation's influence, highlighting the multiple cross-border interaction networks and power systems. These transworld examples are all fuelling a growing global seed sovereignty movement that both challenges domestic states to address the erosion of public seed systems and privatisation, whilst also seeing an increased role for the state, as appropriate governance structures are required and being advanced in many countries, in the face of environmental concerns for biodiversity loss, for seed heritage, for restoration and for future-proofing against climatic change.

Most transformationalists writing in the seed space point to the 'one-size-fits-all' seed policy so characteristic of market globalisation, as being impossible in many agrarian systems providing examples of existing sovereign seed systems, some of which are being enhanced by global forces. They point to ambiguities in new laws sweeping across poorer countries, suggesting that they may never be successfully applied, as:

> There is not one ideal sui generis system that will suit the needs of all countries ... a strong private sector requires a robust public sector to keep options available and to focus on the needs of marginal areas. (IPGRI cited in Tansey 2011, p. 117)

Even controversial new legislative changes leave room for such ambiguities to occur at national, regional and local level, suggesting that such domestic interpretations are more likely in areas where agrarian systems are still largely informal. Sub-Saharan Africa provides the clearest example. Tansey sees the need, and possibility, for a wide range of alternatives, pointing out that there is no obligation for WTO member states to join the UPOV standards, for example, particularly those countries with large informal networks who also want to protect Farmers' Rights (Tansey 2011, p. 117). This position is echoed in the seed literature on African seed systems by De Jonge (2014); Munyi et al. (2016) and Louwaars et al. (2013). For transformationalists, this is an opportunity through the revision of the UPOV Convention (Dutfield 2011, p. 19) for countries like Kenya, for example, to help shape a new system that better reflects these varied farming systems, and the complexity of present international systems that recognises the rights of farmers (Tansey 2011, p. 117). For transformationalists, understanding the variability and availability of choice in seed selection for certain crop varieties is critical in understanding "underlying political configurations and highlighting opportunities for reshaping the terms of the debate particularly under variable conditions" (Scoones and Thompson 2011, p. 1).

In short, there is variation in seed sovereignty. An inherent dynamic now exists which reveals 'distinct local geographies' (Kloppenburg 2014, p. 24), where the use of 'local seeds' has disruptive potentiality for the "dominant model of standardisation and corporate control" (Da Via 2012, p. 238), on the one hand and a variable application of globalising seed agendas across and within countries on the other.

International Organisations

Unlike the hyperglobalist interpretation that IOs are now controlling seed politics everywhere, or the sceptics who say that the state remains the primary actor, these transformationalists point to the flexibility of different countries' interpretations of TRIPS, highlighting that TRIPS allow for variation and that there is scope for different countries to shape their own plant variety regimes and seed laws and policies to suit their development objectives (Tansey 2011, pp. 114–117; Munyi and De Jonge 2015).

The main IOs identified by transformationalist authors who are opening up 'the field of action' and increasing the tensions within institutions and policy agendas are key IOs such as the Convention on Biological Diversity 1992 (CBD), ITPGRFA 1978, 2001 (enforced 2004), UNMDGs, specifically Goal 7, Convention B on biodiversity and poverty (UN 2000),[38] IAASTD 2009 and the Nagoya Protocol 2010. Specifically, the UN Special Rapporteur on Food (De Schutter 2009) has been vehemently outspoken against corporate incursions into subsistence farmers' dynamic seed systems, reflecting intense unease within UN institutions of the impact of global forces.

The emergence of recent international agreements directly challenges IP rules and its property rights system, epitomised by WTO, World Intellectual Property Organisation (WIPO) and TRIPS, which reflect some ambiguities themselves regarding applications of seed practice. The alternative narrative emerging from newer IOs has shifted the emphasis onto concepts such as participatory plant breeding, farmers rights, conservation through use and nature and biodiversity, which themselves represent the "quintessential open source system of innovation" (Tansey 2011, p. 118). All of this creates the dynamic interplay between actors and interests, as well as signalling the crisis of legitimacy and governance now being played out in the seed space, as the varying application of new rules in different countries suggests a multiplicity of geographic effect and intent.

---

[38] http://www.un.org/millenniumgoals/environ.shtml

This offers a lens through which to view the extent to which actors, be they individual states, small farmers, enterprises or other key actors can retain sovereignty over their seed choices. Transformationalists highlight that this varied application of supposedly globalising seed law is already happening in many places. They highlight indications of counter-hegemony within key international institutions and in new agreements and laws, directly relating to seed practices, promoting distinctly alternative paradigms to what has been the conventional modernising orthodoxy in seed/agricultural development models with major IOs for the past few decades.

Transnational Actors
Given the significant shifts that are happening within key IOs, transnational actors and movements are increasing and intensifying within the seed and food space as a battle for control of the future of seed and agricultural systems grows. Since the 1990s, LVC, GRAIN, ETC, Navdanya, Greenpeace, ABN and a plethora of grassroots seed/food initiatives have become key transnational actors in championing and mobilising peasant and small farmer and seed organisations, which are dedicated to agroecological farming methods and food and seed sovereignty. In line with transformationalist thinking, which regards the multidimensional nature of responses to globalisation, they extol the virtues of seed/agrarian systems based on participation, repossession and recovery as examples of a transformative reimagined system of societal organisation that global seed incursions by other powerful transnational actors has unleashed. These new patterns are emerging out of a new kind of seed politics, be they urban activists guerrilla gardening in California, the slow food movement which has emerged in Italy, new seed banks emerging across Europe, Kokopelli (French seedsavers), L'Arche Noah (Austrian seedsavers), ISSA (Irish seedsavers), Heritage seed library (UK), with growing campaigns directly against Monsanto, for GMO-free zones and GM labelling, and for food and seed sovereignty to be enshrined in law all are part of a new transformative seed politics.

Many authors highlight the increased influence of the philanthrocapitalists or TNCs in shaping future narratives, particularly policy agendas, for seed access, use, control, ownership and distribution, especially in vulnerable zones. But they also highlight that most food is still grown by the world's (1.2 billion) poorest subsistence farmers and fisherfolk, who predominantly feed the planet (ETC 2009), largely in heterogeneous, agrobiodiverse organic seed systems. They argue that there is growing resistance to the dominant 'development' model for corporate seed in the form of seed sovereignty movements and initiatives. These TNAs are increasingly targeting the food and seed system, pushing for policy change to protect and promote agrobiodiversity and achieving some success at an institutional level with increasing emphasis on the importance of retaining and recovering genetic diversity and 'farmer seeds' (Tansey 2011; Scoones and Thompson 2011; McCann 2011; IAASTD 2009). The success in achieving legislative addendums, designed to protect seed sovereignty in some of these 'distinct local geographies', suggests evidence that the multidirectional possibilities, central to transformationalist thinking, does indeed lie in the 'field of action' where the contradictory forces of globalisation are most evident. There is a strong normative thrust to "recast globalisation as a 'double-sided process'—or a process of double democratisation" (Held and McGrew 2007, p. 195) and seed provides a good example.

Kloppenburg, clearly a transformationalist, promotes just such a contradictory transworld initiative, when he argues for the development of open source seed access, similar to other open source initiatives in the technology space. His idea is akin to "appropriating the tools of generic globalisation" as Evans had suggested (Evans 2008). Using the 'tools of the master', (i.e. the legal root of state apparatus), enabling repossession, creating biological open sources and the recovery of seed sovereignty (Kloppenburg 2010, 2014) new commons-like spaces would re-emerge.

Seed sovereignty has thrown up multidirectional possibilities, with a strong tendency towards counter-hegemonic globalisation, which is finding traction transworld within and outside of state, IOs, TNCs and NGOs. The emergence of new seed sovereignty movements, some quiet as in Russia (Visser et al. 2015), some vocal and vibrant as in Latin America, in parts of Europe and USA, and some countries in Africa, who are resisting GM such as ABN and AFSA, all signal a complex and divergent application of seed laws and policies globally. Raj Patel had referred to these 'sovereign spaces' as "overlapping sovereignties and jurisdictions" where food sovereignty "is an invitation to reimagine the very notion of political constituency along overlapping ecological lines".[39]

---

[39] Chandra Kumar, 'The Politics of Starving: An Interview with Raj Patel', available at http://uppingtheanti.org/journal/article/11-the-politics-of-starving. Accessed 26 December 2018.

CONCLUSION

This chapter has traced the changing discourse on food security and its implications for seed sovereignty since the middle of the twentieth century. It examined what hyperglobalism, scepticism and transformationalism say about the effects of globalisation on seed sovereignty in three key areas, the domestic state, international organisations and transnational actors. In the next chapter, I will identify how I intend to test the implications of the three schools of thought with the aim of identifying which, if any, of the approaches best captures what is happening in relation to the practice of seed sovereignty.

REFERENCES

ACB. 2015. *The expansion of the commercial seed sector in sub-Saharan Africa: major players, key issues and trends* [Online]. Available from: www.acbio.org.za [Accessed 3rd February 2016].

ACB. 2017. *The Bayer-Monsanto merger: implications for South Africa's agricultural future and its smallholder farmers* [Online]. Available from: http://acbio.org.za/wp-content/uploads/2017/02/Bayer-Monsanto-report.pdf [Accessed 4th March 2017].

Acharya, A. 2011. Human security. *IN:* Baylis, J., Smith, S. and Owens, P. (eds.) *The globalization of world politics, an introduction to international relations.* 5th ed. New York: Oxford University Press, pp. 478–493.

AFSA and GRAIN. 2015. *Land and seed laws under attack, who is pushing changes in Africa?* [Online]. Available from: http://www.grain.org [Accessed 22nd January 2016].

Alden, C. 2007. *China in Africa.* London: Zed Books.

Alemu, D. 2011. The political economy of Ethiopian cereal seed systems: state control, market liberalisation and decentralisation. *IDS Bulletin,* 42(4), pp. 69–77.

Altieri, M.A. 2009. Agroecology, small farms and food sovereignty. *Monthly Review* [Online]. Available from: http://search.proquest.com/openview/981 65d33d3bec3c0f47ca6821e49eef0/1, 61(3), pp. 102–113 [Accessed 6th December 2010].

Amanor, K.S. 2011. From farmer participation to pro-poor seed markets: the political economy of commercial cereal seed networks in Ghana. *IDS Bulletin,* 42(4), pp. 48–58.

Berne Declaration. 2013. *Agropoly: a handful of corporations control world food production.* Zurich, Switzerland: Berne Declaration & EcoNexus.

Berne Declaration. 2014. *Owning seeds, accessing food, a human rights impact assessment of UPOV 1991. Based on case studies in Kenya, Peru and the Philippines.* Zurich, Switzerland: Berne Declaration.

Carmody, P. 2011. *The new scramble for Africa*. Cambridge: Polity.

Carolan, M. 2012. *The sociology of food and agriculture*. Earthscan Routledge: New York.

Cerny, P. 2010. *Rethinking world politics, a theory of transnational neopluralism*. Oxford: Oxford University press.

Chinsinga, B. 2011. Seeds and subsidies: the political economy of input programmes in Malawi. *IDS Bulletin*, 42(4), pp. 59–68.

Clapp, J. 2014. Food security and food sovereignty: getting past the binary. *Dialogues in Human Geography*, 4(2) 206–211.

Coordination Sud. 2017. *The right to seeds: a fundamental right for small farmers!* Paris, France: Coordination Sud.

Da Via, E. 2012. Seed diversity, farmers' rights, and the politics of repeasantisation. *International Journal of Sociology of Agriculture and Food*, 19 (2), pp. 229–242.

De Jonge, B. 2014. Plant variety protection in sub-Saharan Africa: balancing commercial and smallholder farmers' interests. *Journal of Politics and Law*, 7 (3), pp. 100–111.

De Schutter, O. 2009. *Seed policies and the right to food: enhancing agrobiodiversity and encouraging innovation report of the Special Rapporteur*. A/64/170 United Nations General Assembly [Online]. Available from: http://farmersrights. org/pdf/righttofood-n0942473.pdf [Accessed 23rd February 2013].

Downes, G. 2003. *Implications of TRIPS for food security in the majority world* [Online]. Available from: http://comhlamh.org/wp-content/uploads/2013/ 09/Implications-of-Trips-for-Food-Security.pdf [Accessed 10th April 2017].

Dutfield, G. 2011. Food, biological diversity and intellectual property: the role of the International Union for the Protection of New Varieties of Plants (UPOV). *Intellectual Property Issue*, Paper No.9. Quaker United Nations Office.

Edelman, M. 2014. Food sovereignty: forgotten genealogies and future regulatory challenges. *Journal of Peasant Studies*, 41(6), pp. 959–978.

Esquinas-Alcazar, J. 2005. Protecting crop genetic diversity for food security: political, ethical and technical challenges. *Nature*, 6, pp. 946–953, www.nature. com/reviews/genetics.

ETC (Erosion, Technology and Concentration Group). 2009. *Who will feed us? Questions for the food and climate crises* [Online]. Available from: http://www. etcgroup.org/sites/www.etcgroup.org/files/web_who_will_feed_us_with_ notes_0.pdf [Accessed 1stFebruary 2012].

ETC (Erosion, Technology and Concentration Group). 2010. *Capturing climate genes. Gene giants stockpile 'climate-ready' patents* [Online]. Available from: http://www.etcgroup.org/content/gene-giants-stockpile-patents- "climateready"-crops-bid-become-biomassters-0 [Accessed 1st February 2012].

Evans, P. 2008. Is an alternative globalisation possible? *Politics and Society*, 36(271), pp. 271–305.

FAO. 1983. *World food security: a reappraisal of the concepts and approaches*. Director General's Report: Rome. Food and Agricultural Organisation of the United Nations.

FAO. 1996. *Rome Declaration on world food security and World Food Summit Plan of Action.* World Food Summit, Rome 13–17 November. Rome: Food and Agriculture Organisation of the United Nations.

FAO. 2003. *Trade reforms and food security: conceptualising the linkages.* Rome: Food and Agricultural Organisation of the United Nations.

FAO. 2006. Food Security. *FAO Policy Brief,* June 2006, Issue 2. Available at http://www.fao.org/fileadmin/templates/faoitaly/documents/pdf/pdf_Food_Security_Cocept_Note.pdf. Accessed 26 Dec 2018.

FAO. 2009. *Proceedings of the second World Seed Conference.* Responding to the challenges of a changing world: The role of new plant varieties and high quality seed in agriculture. FAO Headquarters, Rome, September 8–10 [Online]. Available from: www.fao.org/docrep/014/am490e/am490e00.pdf [Accessed 11th May 2013]. Rome: FAO.

FAO. 2010. *The second report on the state of the world's plant genetic resources for food and agriculture* [Online]. Available from: www.fao.org/docrep/013/i1500e/i1500e.pdf [Accessed 19th December 2015] Rome: FAO.

FAO. 2011. *Potential effects of climate change on crop pollination.* Rome: FAO.

FAO. 2014. *The state of food and agriculture. Innovation in family farming* [Online]. Available from: www.fao.org/publications/sofa/2014/en [Accessed 11th May 2013]. Rome: FAO.

FAO. 2015. *The state of food insecurity in the world.* Rome: FAO.

FAO. 2016. *Climate change and food security: risks and responses.* Rome: FAO.

Focus on the Global South and GRAIN. 2004. *Iraq's new patent law: a declaration of war against farmers* [Online]. Available from: https://www.grain.org/article/entries/150-iraq-s-new-patent-law-a-declaration-of-war-against-farmers [Accessed 31st March 2017].

Future Agricultures. 2014. Emerging seed markets: the role of Brazilian, Chinese and Indian seeds in African agriculture. *Future Agricultures* Policy Brief 79, October 2014 [Online]. Available from: www.future-agriculture.org. [Accessed 3rd August 2015].

Gates Foundation and AGRA. 2015. *Investing in agriculture to reduce poverty and hunger* [Online]. Available from: http://www.gatesfoundation.org/How-We-Work/Resources/Grantee-Profiles/Grantee-Profile-Alliance-for-a-Green-Revolution-in-Africa-AGRA [Accessed 21stOctober 2015].

George, S. 2015. *Shadow sovereigns—how global corporations are seizing power.* Cambridge UK: Polity Press.

Gilpin, R. 2001. *Global political economy.* Princeton University Press: Princeton.

GRAIN. 2005. *Africa's seed laws: red carpet for the corporations* [Online]. Available from: https://www.grain.org/article/entries/540-africa-s-seeds-laws-red-carpet-for-corporations [Accessed 11th November 2014].

Grey, S. and Patel, R. 2015. Food sovereignty as decolonisation: some contributions from indigenous movements to food systems and development politics. *Agriculture and Human Values* (32) pp. 431–444.

Harrison, G. 2010. *Neoliberal Africa: The impact of global social engineering.* London Zed Books.

Harvey, D. 2005. *A brief history of neoliberalism.* Oxford: Oxford University Press.

Hassan, G. 2005. *Biopiracy and GMOs: the fate of Iraq's agriculture* [Online]. Available from: http://www.globalresearch.ca/biopiracy-and-gmos-the-fate-of-iraq-s-agriculture/1447 [Accessed 12th March 2013].

Hay, C. and Marsh, D. 2000. *Demystifying globalization.* London and New York: Macmillan Press.

Held, D. and McGrew, A. 2007. *Globalisation/anti-globalisation: beyond the great divide.* 2nd ed, Cambridge: Polity Press.

Hettne, B. 2009. *Thinking about development—development matters.* London: Zed Books.

Holt Gimenez, E. and Shattuck, A. 2011. Food crises, food regimes and food movements: rumblings of reform or tides of transformation? *Journal of Peasant Studies,* 38 (1), pp. 109–144.

IAASTD. 2009. *Agriculture at a crossroads: a synthesis of the global and sub-global IAASTD reports.* Washington, DC: Island Press.

IPCC. 2014. *Climate change 2014: synthesis report, contribution of working groups I, II and III to the fifth assessment report of the Intergovernmental Panel on Climate Change* [Online]. Available from: http://ipcc.ch/pdf/assessment-report/ar5/syr/AR5_SYR_FINAL_SPM.pdf. [Accessed 30th March 2017].

Jarosz, L. 2014. Comparing food security and food sovereignty discourses. *Dialogues in Human Geography,* 4(2) pp. 168–181.

Jessop, B. 1997. Capitalism and its future: remarks on regulation, government and governance. *Review of International Political Economy,* 4(3), pp. 561–581.

Kloppenburg, J. 2008. *Seeds, sovereignty, and the Via Campesina. Plants, Property, and the Promise of Open Source Biology* [Online]. Available from: https://www.researchgate.net/publication/255583305_Seeds_Sovereignty_and_the_Via_Campesina_Plants_Property_and_the_Promise_of_Open_Source_Biology [Accessed 10th April 2017].

Kloppenburg, J. 2010. Impeding dispossession, enabling repossession: biological open source and the recovery of seed sovereignty. *Journal of Agrarian Change,* 10(3), pp. 367–388.

Kloppenburg, J., 2014. Re-purposing the master's tools: the open source seed initiative and the struggle for seed sovereignty. *Journal of Peasant Studies,* 41(6), pp. 1225–1246.

Korab-Karpowicz, W.J. 2010. Political realism in international relations. *Stanford Encyclopedia of Philosophy* [Online]. Available from: https://plato.stanford.edu/entries/realism-intl-relations/. [Accessed 28th April 2014]

Kumashiro, K. 2012. When billionaires become educational experts. *Academe,* 99(3), pp. 10–16.

Lahiff, E. 2007. Willing buyer, willing seller: South Africa's failed experiment in market-led agrarian reform. *Third World Quarterly,* 28(8), pp. 1577–1597.

Lahiff, E., Borras, S M, and Kay C. 2007. Market-Led Agrarian Reform: Policies, Performance and Prospects. *Third World Quarterly,* 28 (8), pp. 1417–1436.

Lamy, S.L. 2011. Contemporary mainstream approaches: neo-realism and neo-liberalism. *IN:* Baylis, J., Smith, S. and Owens, P. (eds.) *The globalization of world politics, an introduction to international relations.* 5th ed. New York: Oxford University Press, pp. 114–129.

Langyintuo, A. 2011. *African agriculture and productivity.* Nairobi, Kenya: Alliance for a Green Revolution.

La Via Campesina (LVC). 1996. *The right to produce and access to land* [Online]. Available from: http://www.acordinternational.org/silo/files/decfoodsov1996.pdf [Accessed 22nd December 2018].

La Via Campesina (LVC). 2008. Declaration of Maputo. *IN: V International Conference of LVC, October 19–22* [Online]. Available from: http://viacampesina.org/en/index.php/our-conferences-mainmenu-28/5-maputo-2008-mainmenu-68/declarations-mainmenu-70/602-open-letter-from-maputo-v-international-conference-of-la-vcampesina [Accessed 22nd June 2015].

Louwaars, N.P., De Boef, W.S. and Edeme, J. 2013. Integrated seed sector development in Africa: A basis for seed policy and law. *Journal of Crop Improvement.* 27, pp. 186–214.

LVC. 2001. The position of Vía Campesina on biodiversity, biosafety and genetic resources. *Development,* 44(4), 47–51.

Maathai, W. 2010. *The challenge for Africa.* London: Arrow Books.

Magdoff, F. and Tokar, B. 2010. *Agriculture and food in crisis; conflict, resistance, and renewal.* New York: Monthly Review Press.

McCann, J.C. 2011. The political ecology of cereal seed development in Africa: a history of selection. *IDS Bulletin,* 42(4), pp. 24–35.

McGuire, S. and Sperling, L. 2011. The links between food security and seed security: facts and fiction that guide response. *Development in Practice.* 21(4–5), pp. 493–508.

McKeon, N. 2015. *Food security governance; empowering communities, regulating corporations.* London and New York: Routledge, Taylor and Francis.

McMichael, P. and Schneider, M. 2011. Food security politics and the millennium development goals. *Third World Quarterly,* 32(1), pp. 119–139.

Minde, I. 2004. *Harmonizing seed policies and regulations in Eastern Africa: experiences and lessons learned.* Entebbe, Uganda: Eastern and Central African Programme for Agricultural Policy Analysis.

Mooney, P. 1979. *Seeds of the earth: a private or public resource.* Inter Pares for the Canadian Council for International Co-operation and The International Coalition for Development Action.

Mooney, P. 1983. The law of the seed—another development and plant genetic resources. *Development Dialogue,* 1(2).

Munyi, P. 2015. Plant variety protection regime in relation to relevant international obligations: implications for smallholder farmers in Kenya. *The Journal of Intellectual Property,* 18(1–2), pp. 65–85.

Munyi, P. and De Jonge, B. 2015. Seed systems support in Kenya: consideration for an integrated seed sector development approach. *Journal of Sustainable Development*, 8(2), pp. 161.

Munyi, P., De Jonge, B. and Visser, B. 2016. Opportunities and threats to harmonisation of plant breeders' rights in Africa: ARIPO and SADC. *African Journal of International and Comparative Law*, 24(1), pp. 86–104.

Murphy, S. 2013. *Land grabs and fragile food systems—the role of globalisation*. Institute for Agriculture and Trade Policy.

Murphy, S. 2014. Expanding the possibilities for a future free of hunger. *Dialogues in Human Geography* [Online], 4(2), pp. 225–228. Available from: http://journals.sagepub.com/doi/pdf/10.1177/2043820614537166   [Accessed 26th September 2014].

Nabhan, G. 1997. *Cultures of habitat*. Washington, DC: Counterpoint.

New Internationalist. 2010. Seeds—the facts. *New Internationalist*, 435, pp. 10.

Oakland Institute. 2017. *Down on the seed: The World Bank enables corporate takeover of seeds*. California, USA: Oakland Institute.

Odame, H. and Muange, E. 2011. Can agro-dealers deliver the green revolution in Kenya? *IDS Bulletin*, 42(4), pp. 78–89.

Ohmae, K. 1995. *The end of the nation state: the rise of regional economies*. London: Harper Collins Publishers.

Patel, R. 2007. *Stuffed and starved*. London: Portobello Books.

Patel, R. 2009. What does food sovereignty look like? *The Journal of Peasant Studies*, 36(3), pp. 663–706.

Remington, T., Maroko, J., Walsh, S., Omanga, P. and Charles, E. 2002. Getting off the seeds-and-tools treadmill with CRS seed vouchers and fairs. *Disasters*, 26(4), pp. 316–328.

Robin, M. 2010. *The world according to Monsanto: pollution, corruption, and the control of our food supply*. New York: The New Press.

Rosset, P. 2003. Food sovereignty: global rallying cry of farmer movements. *Institute for Food and Development Policy Backgrounder*, 9(4), pp. 1–4.

Scholte, J.A. 2008. Defining globalisation. *The World Economy*, 31(11), pp. 1471–1502.

Scoones, I. and Thompson, J. 2011. The politics of seed in Africa's green revolution. alternative narratives and competing pathways. *IDS Bulletin*, 42(4), pp. 1–23.

Sen, A. 1981. *Poverty and famines*. Oxford: Oxford University Press.

Shaw D. J. 2007. *World Food Security: A History Since 1945*. London: Palgrave Macmillan.

Shiva, V. 1993. *Monocultures of the mind: perspectives on biodiversity and biotechnology*. London and New Jersey: Zed Books

Shiva, V. 2005. *India divided: diversity and democracy under attack*. New York: An Open Media Book, Seven Stories Press.

Shiva, V. 2013. *The law of the seed*. [Online]. Available from: www.navdanya.org/attachments/Latest_Publications4.pdf [Accessed 2nd April 2015].

Sklair, L. 2002. *Globalisation, capitalism and its alternatives*. Oxford: Oxford University Press.

Sperling, L. 2008. *When disaster strikes: a guide for assessing seed security*. Cali: CIAT.

Stiglitz, J. 2005. The overselling of globalisation. *IN* Weinstein, M.M. (ed.), *Globalisation: what's new*. New York: Columbia University Press, pp. 228–262.

Strange, S. 1996. *The retreat of the state*. Cambridge: Cambridge University Press.

Tansey, G. 2011. Whose power to control? Some reflections on seed systems and food security in a changing world. *IDS Bulletin*, 42(4), pp. 111–120.

The Guardian. 2014. G8 New Alliance condemned as a new wave of colonialism in Africa. *The Guardian* [Online], 18th February. Available from: https://www.theguardian.com/global-development/2014/feb/18/g8-new-alliance-condemned-new-colonialism [Accessed 14/8/2014].

Thompson, C.B. 2014. Philanthrocapitalism: appropriation of Africa's genetic wealth. *Review of African Political Economy*, 41(141), pp. 389–405.

UN. 2000. *We the peoples. The role of the United Nations in the 21st century*. New York: United Nations.

UNDP-UNDAF. 2011. *Ethiopia United Nations Development Assistance Framework 2012–2015*. United Nations Country Team, March 2011.

USAID. 1995. *Food aid for food security, policy paper*. Washington, DC: Bureau for Programme and Policy Coordination, USAID.

USAID. 2012. *Ethiopia country development cooperation strategy 2011–2015: accelerating the transformation toward prosperity* [Online]. Available from: https://www.usaid.gov/sites/default/files/documents/1860/Country%20Development%20Cooperation%20Strategy%20-%20Ethiopia%20March%202012.pdf. [Accessed 12th January 2014].

USDA Foreign Agricultural Service. 2008. Kenya planting seed report. GAIN Report number: KE8010.

Visser, O., Mamonova, N., Spoor, M. and Nikulin, A. 2015. 'Quiet food sovereignty' as food sovereignty without a movement? Insights from post-socialist Russia. *Globalizations*, 12(4), pp. 513–528.

Windfuhr, M. and Jansen, J. 2005. *Food sovereignty: towards democracy in localised food systems*. Rugby, Warwickshire: FIAN ITDG Publishing.

World Bank. 1986. *Poverty and hunger*. Washington, DC: World Bank.

World Bank. 2008. *World Bank Development Report: Agriculture for Development*. Washington DC: World Bank.

World Bank. 2013a. *World Bank Group Agriculture Action Plan 2013–1015* [Online]. Available from: http://documents.worldbank.org/curated/en/2013/03/1749873:bank-group-agriculture-action-plan-2013-2015. [Accessed 22nd June 2015].

World Bank. 2013b. *World development report. Risk and opportunity: managing risk for development*. Washington, DC: World Bank.

CHAPTER 4

Kenya: A Hyperglobalised Seed Law

Ethiopia and Kenya are identified in the literature as important areas in the intensifying discourse regarding power, control, risks and benefits (Tansey 2011, p. 117) with regard to changes currently happening in the African seed space. This chapter presents the case of Kenya's Seed and Plant Varieties (Amendment) Act (SPVAA 2012). The key questions are: Who wrote the law? How was it drafted? What was the motivation behind the content of the law? I am looking to identify the key actors involved in bringing the law to fruition? Who was included in the process of determining its contents? Who was excluded from the process? The aim of this exercise is to answer the underlying research question—where does seed sovereignty lie in the face of globalisation? Which approach to globalisation—hyperglobalist, sceptical or transformationalist—best describes how sovereignty is exercised in the making of the new law?

The Seed and Plant Varieties (Amendment) Act was enacted by the Parliament of Kenya on 4 January 2013. SPVAA 2012 marked the first amendment of a Kenyan seed law in a decade.[1] It is significant as it is the first seed law following, first, the TRIPS Agreement of 2002, second, the establishment of AGRA, an alliance between the giant transnational 'philanthrocapitalist' foundations, the Rockefeller Foundation and BMGF

---

[1] The Seed and Plant Varieties Act, 1972 (as amended in 2002) Act No. 2 of 2002 (Cap 326).

© The Author(s) 2019
C. O'Grady Walshe, *Globalisation and Seed Sovereignty in Sub-Saharan Africa*, International Political Economy Series, https://doi.org/10.1007/978-3-030-12870-8_4

117

in 2006, with headquarters in Nairobi, and, third, the enactment of a new Kenyan seed policy (2010).[2] These three events encompass the three main categories of actor previously identified, namely the international organisations, transnational actors and domestic/state actors. An examination of Kenya, a recognised pivotal interlocutor for the Global South in general and the African continent specifically, provides a good test of the premise underlying the different perspectives on globalisation in this period, allowing us to assess the role and motivations of the different actors in determining outcomes affecting seed sovereignty.

This chapter draws upon the extensive unstructured interviews with all the relevant seed actors which were conducted during fieldwork in Kenya. These interviews combined with extensive data sourcing of all available legal and policy documents, parliamentary records and related material provide the detailed substance of this chapter. In the Kenyan case much of this documentation was available online at Hansard online. This was an invaluable source of information, allowing immediate comparative analysis of the relevant seed laws, noting principal changes between SPVAA 2012 and the preceding law CAP 326 from 2002. This provided a crucial temporal boundary for the study and clear examples of the changes enacted. A wider literature review, particularly the Institute of Development Studies (IDS) Special edition "The Politics of Seed in Africa's Green Revolution" (2011) was informative and signposted further in-depth online searches of other important actors who had a role in the seed space at this time in Kenya. The establishment of AGRA in 2006 was a key event. It provided much press and media coverage, which in turn revealed other actors and important events that shaped the making of this law. This method of document collection and analysis was employed throughout and constantly informed the construction of the detailed roadmap which emerged from this study. Similarly, contact with NGOs was vital throughout. Despite being excluded from the process of law making in this case, they were extremely engaged and offered critical information which filled in many gaps in this case. The engagement with so many diverse interviewees provided invaluable local knowledge and essential background information, ensuring reflexivity throughout.

This chapter has three sections. The first section identifies and examines the most salient legislative changes which occurred with the passage of SPVAA 2012. The second section lays out the key fora where legislative changes were drafted, and details the legislative and regulatory road map

---

[2] Republic of Kenya, Ministry of Agriculture, National Seed Policy, June 2010.

which led to the changes enacted. The final section provides a detailed analysis of the main internal and external actors involved in the decision-making process and assesses their role and motivations in the crucial years which culminated in the passage of the new Seed Law.

SPVAA 2012: A New Seed Act

Kenya's most recent seed law (SPVAA 2012) is an amended successor to previous seed and plant varieties legislation, notably the official Kenyan Seed and Plant Varieties Act (1972)[3] and the 2002 amended version of that law known as Chapter 326—Seed and Plant Varieties Act, otherwise known as (Cap 326).[4] It is important to note that the preparation of SPVAA 2012 occurred concurrently with the 2010 revision of the Kenyan Constitution, and a suite of other relevant new domestic laws pertaining to changing agricultural practices, namely the Crops Act (2013),[5] the Agriculture, Fisheries and Food Authority (AFFA Act) (2013),[6] the Kenyan Agricultural and Livestock Research Act (2013),[7] the Kenya Plant Health Inspectorate Service (Kephis) Act 2012,[8] the Biosafety Act (2009)[9] and the Anti-Counterfeit Act (2008).[10] However, SPVAA 2012 is the predominant domestic seed law that deals exclusively with the issues which directly affect seed sovereignty (namely farmers' rights, plant-breeders' rights [PBRs], plant variety protection [PVP] and seed certification rules). There are also a number of important international and regional laws and treaties which act as key drivers of the changing juridical field in Kenya and beyond. However, I am concentrating on the 2012 Act.

To introduce the changes brought about by SPVAA 2012, I have selected seven aspects of the 2012 law that are markedly different from previous Kenyan seed laws.

---

[3] The Seed and Plant Varieties Act (Act No. 1 of 1972, L.N. 152/1998, Act No. 2 of 2002 (Cap 326), Act No. 53 of 2012, L.N. 71/2011).

[4] The Seed and Plant Varieties Act, 1972 (as amended in 2002).

[5] The Crops Act (Act No. 16 of 2013, L.N. 57/2013, L.N. 110/2014).

[6] The Agricultural, Fisheries and Food Authorities (Act, No. 37 of 2013).

[7] The Kenyan Agricultural and Livestock Research Act, 2013.

[8] The Kenya Plant Health Inspectorate Act, 2012 (Kenya Gazette Supplement No. 218 (Acts No. 54)).

[9] The Biosafety Act (Act No. 2 of 2009, Cap 321 A).

[10] The Anti-Counterfeit Act (2008) (relevant regarding infringements of PBRs).

1. Harmonisation: UPOV 91

The SPVAA 2012, in amending the Principal Act of Section 3 of Cap 326, calls for "the integration and harmonisation of the seed industry" (Section 4 b) (i) (j). In the Memorandum of Objects and Reasons proposing the SPVAA Bill 2011,[11] Clauses 15, 16, 17, 18 and 26, which were enacted in SPVAA 2012, are all cited as the necessary changes to Cap 326 to ensure compliance with UPOV 91. This is the language of the agribusiness corporate paradigm. It marks a significant shift in the direction which UPOV 91 and other international institutions, such as TRIPS, WTO, AGRA, OECD and intellectual property regimes such as UN World Intellectual Property Organisation (WIPO) and its regional counterpart, the African Regional Intellectual Property Organisation (ARIPO) and corporate bodies had been calling for. UPOV 91 compliance marks a regime change characterised by PBRs, PVP and Intellectual Property Rights (IPRs). The impact of this shift towards transglobal governance over domestic seed systems is central to this case study and our understanding of the changing nature of sovereignties underlying what Rangnekar (2014, p. 361) calls this increasingly "polycentric fragmented global IP architecture".

2. Stricter PBRs

There are whole sections of SPVAA 2012 Act, notably Section 16 and Section 17, which relate to changes in PBRs and enshrine much greater proprietorial advantage for the commercialisation of seed and plant varieties named, claimed and labelled by a plant breeder. SPVAA 2012 Section 14 amends Section 17 of the Principal Act on two counts (Cap 326, Part V, 'Plant-Breeders' Rights'). Firstly, it deletes the reference to ministerial discrimination in determining which plant varieties and species may be granted PBRs, substituting the words "varieties of all plant genera and species". It thereby establishes much more latitude for breeders' rights across the entire plant/seed spectrum. Secondly, it specifically deletes sub-section 4 of the same Section 17 regarding PBRs, which had been the only part of the principal act which mentioned consultation with representatives of such organisations the "Minister deems to have substantial interest in the matter to be

---

[11] Memorandum of Objects and Reasons to the Seeds and Plant Varieties (Amendment) Bill, 2011, dated 28 November 2011.

regulated". This would have allowed space for consultation with domestic farmers and civil society organisations on matters which greatly affect their rights and livelihoods. Clauses 15, 16, 17, 18 and 19 of SPVAA Bill of 2011 pertaining to PBRs under UPOV rules were all enacted in SPVAA 2012. Clause 24 and 25 establish a penalty system for contravention of the new SPV rules in the 2012 Act.

3. Strict PVP Rules Certification and Related Aspects

Sections 16 and 17 of SPVAA 2012 lay out much more clearly what plant variety protection means under Kenya's new Seed Act. The protected variety actually means protection for the breeder. To this end, Section 17 lays out the conditions of permission for production and reproduction, propagation, selling or other marketing, exporting and importing, an essential part of complying with the strictest interpretation (i.e. internationally in UPOV 91 and at regional PVP legal systems through ARIPO, Arusha Protocol, South African Development Community [SADC], Organisation Africaine de la Propriete Intellectuelle [OAPI] and Common Market for Eastern and Southern Africa [COMESA] of PBRs and PVP [SPVAA 2012, Section 17 (a) (ii) a–g]).

Specifically at Section 17.1(E) SPVAA 2012 states in relation to farmers' rights in this instance of commercialised seed/crops:

> Within reasonable limits and subject to the safeguarding of the legitimate interests of the breeder, farmers may use the product of the harvest which they have obtained by planting, on their own holdings, the protected variety.

This is significant in that it brings Kenyan Seed and plant variety law in line with the scope of rights afforded under strict international UPOV 91 law (despite having no obligation under international law to do so),[12] which delivers a wide breadth of breeders' rights, while farmers' rights are

---

[12] Firstly, WTO Agreement on TRIPS (TRIPS, Article 27.3 (b)) to which Kenya is a party requires member states to provide IP protection for plant varieties, but allows governments a wide latitude in its determination (Dutfield 2011, p. 7). Secondly, Kenya is already a member of UPOV 78, which allows wider scope regarding PBRs and is considered a better option for African countries than the stricter UPOV 91, which was designed with 'developed' agricultural systems in mind. Despite SPVAA 2012 being compliant with UPOV 91, Kenya has yet to deposit an instrument of accession (Munyi and De Jonge 2015) and so is still at time of writing bound only by UPOV 78.

greatly diminished. Specifically, it is noted by Munyi and De Jonge (2015) that:

> remarkably, Kenya's 2012 SPVA Amendment, which was implemented with the aim to make Kenya's PBR law compliant with UPOV 1991 does not include the private and non-commercial use exemption as required by UPOV. (Munyi and de Jonge 2015, p. 170)

Thus, "the new law leaves little room for small-holder farmers to be allowed to freely use farm-saved seed of protected varieties" (ibid.). All the core features of seed sovereignty as defined previously, such as seed saving and exchanging across family and community, the reuse of farm-saved seed and maintenance of informal supply and distribution channels, which is widely practiced throughout Kenya's vast subsistence farming population for many seed/crop varieties and which was previously allowable with no restrictions will now be subject to explicit provisions and conditions. This part of SPVAA 2012 puts plant-breeders' rights directly in conflict with farmers' rights, and unsurprisingly was the subject of protracted dispute, not least because of its contradiction with the newly agreed National Constitution 2010 (Article 11 3 (b)).[13]

4. "Agroecological" Deleted

Section 6 of SPVAA 2012 amends Section 8 of Cap 326 and specifically deletes the words 'agroecological value', substituting the words 'value for cultivation and use' instead. This is significant as the term 'agroecological' is closely associated with the language of farmers' rights and the seed and food sovereignty paradigm, as well as other key international organisations in the transglobal domain such as Convention on Biological Diversity (CBD), International Treaty on Plant Genetic Resources for Food and Agriculture (ITPGRFA) of Food and Agriculture Organisation (FAO), Nagoya Protocol, International Assessment of Agricultural Knowledge, Science and Technology for Development (IAASTD), all of which press for such approaches to be considered. Therefore, the decision to delete the term at this time is significant for the purposes of this study.

---

[13] 11 (3) (b) provides that "Parliament shall enact legislation to recognise and protect the ownership of indigenous seeds and plant varieties, their genetic and diverse characteristics, and their use by the communities of Kenya" (The Constitution of Kenya 2010).

5. Caters for Emerging Seed Technologies

Section 13 of SPVAA 2012 amends Section 16 of the Principal Act and includes a new sub-section which specifically states: "The Service may, with the approval of the Minister by order in the Gazette (a) develop guidelines and protocols for the management of emerging technologies in seed production" Section 13 2 (a) SPVAA 2012, while a new sub-section 2 (b) goes on to state

> and apply this section to any area in Kenya where persons are engaged in growing crops for seeds of any type or variety of plant specified in the order, if he is satisfied that in that area satisfactory arrangements have been made for locating such crops so as to isolate them from crops or plants which might cause injurious cross-pollination. (SPVAA 2012, Section 13)

This opens the door to allow 'novel' technologies, such as genetically engineered seeds or corporate seeds, and makes way for their easier passage in Kenya. This is noteworthy as it corresponds to the highly contested Biosafety Act (2009) that had recently been passed by the Kenyan Parliament which specifically was established "to regulate the activities in relation to GMOs", to facilitate their use (Biosafety Act 2009, p. 85). This is at variance with the rhetoric of the Kenyan state at other international fora, where the country has lobbied strongly against 'patents on life' and for the African Model Law at TRIPS Council (Rangnekar 2014), yet where it has done the exact opposite in its domestic laws.

6. PGRs: Open to Commercial/Research Interests

SPVAA 2012 Section 20 inserts a whole new section at 27A of Cap. 326 establishing a "National Plant Genetic Resources Centre (NPGRC) which shall be responsible for the conservation and sustainable utilisation of plant biodiversity in Kenya". Though loosely defined and with no mechanism established to activate it, some of the functions listed for the establishment of the NPGRC may be contradictory in nature, as they pertain not only to the protection of biodiversity and rights of ownership over indigenous varieties, but also stipulate the need to "cooperate with international institutions", "including the administration of material transfer arrangements" (27A.2. (d)) and to "collaborate with institutions of higher learning to address adaptive, applied and strategic research" (SPVAA

2012.27 A.2. (g)), which in the context of controversial funding by agrochemical seed corporations and philanthrocapitalist interest groups of
certain universities as well as their own vested interest in developing GM
crops[14] may compromise key elements of PGR conservation in this case.

7. Kephis

Powers are handed over in this amending Act to Kephis, a government
parastatal agency with responsibility for regulating seed in Kenya. Referred
to as the 'Service' throughout the new Act, Part 5 of SPVAA 2012 inserts:

> The Service shall be the national designated authority (NDA) for matters
> relating to seeds and plant variety protection and shall, subject to the direc-
> tions of the Minister, be responsible for the administration of this Act.
> (Section 5:3A SPVAA 2012)

There are specific sections where the word 'Minister' is deleted in
SPVAA 2012 and substituted by 'the service' (Sections 10,12, 13 amending Sections 11, 15, 16 of Cap 326).

The story of Kephis is important as it was first set up when the seed
industry was liberalised in 1996 in Kenya. Now, it is being given more
power through SPVAA 2012 to enforce the Act, yet simultaneously some
of its functions can be privatised, where for the "purposes of enforcing
the Act, the Service may authorise competent private or public persons to
perform specified functions under this Act on its behalf" (SPVAA 2012,
Section 5 3B 1 (a) (b)). In this way, Kephis, whilst being promoted as the
NDA, is nonetheless also being divested of power, signalling a loosening
of direct state involvement in key seed regulatory functions, something
which was keenly sought by private-sector interests (USDA 2008; STAK
2007). The establishment of Kephis as the 'designated authority' to over-
see the implementation and administration of these stricter seed controls
of PBRs, PVP and enforcement of the new national biosafety regulations
was further secured through a new Kephis Act which passed through

---

[14] The Rockefeller Foundation is listed as a key institution involved in the first confined
field research trial in Kenya for maize crop (*Zea Mays* 1, Cry 1Ba) alongside Monsanto,
Syngenta, International Maize and Wheat Improvement Centre (CIMMYT), University of
Ottawa and KARI (Karembu et al. 2009, p. 12).

parliament on the same day as SPVAA 2012.[15] This copper fastened the essential ingredients of a new Kenyan seed regime which was achieved through SPVAA 2012.

Overall, it is clear that SPVAA 2012, as enacted in January 2013, marked a critical moment of institutional and legal change for the Kenyan State, altering the power structure and sovereign control of the country's seed sector in favour of liberalisation and increased private-sector engagement. It has created a more porous-enabling environment, facilitating commercial intervention in opening up seed markets, and has introduced globalising legal instruments which inevitably affect the practice of seed sovereignty as constituted by the majority 80% of the smallholder farming population who rely on complex, highly evolved 'farmer-managed' (ACB 2015) seed systems.

In the next section I detail the exact chronology of events and identify the actors involved in the years preceding the enactment of the new seed law and highlight the key moments in the process where the changes listed above occurred. The actors emerged both through the study of the legislative/parliamentary process, and by way of the comprehensive list of people and organisations involved in the formulation of the National Seed Policy Committee which preceded the new seed law. Here, representative organisations for the seed TNCs emerged as central actors, leading to further investigation of their activities and documentation online. Cross referencing with NGOs on the ground and the uncovering of Seed Trade Association of Kenya (STAK) memos from that time was invaluable. This identified specific interventions, meetings and events which were instrumental in formulating the new law. The most noteworthy of these was confirmation regarding STAK's inclusion and influence in two task forces which were established to change the seed law and seed policy between 2005 and 2006. This provided detailed information of the personnel involved and the critical changes sought and achieved from their perspective. This led me to further investigate the role of notable representatives and bodies such as COMESA, Association for Strengthening Agricultural Research in Eastern and Central Africa (ASARECA), Eastern and Central Africa Programme for Agricultural Policy Analysis (ECAPAPA) and other forums where important decisions were made. The comprehensive roadmap which follows is informed by drawing together all of this documented evidence.

---

[15] The Kenya Plant Health Inspectorate Act, 2012 (Kenya Gazette Supplement No. 218 (Acts No. 54)).

The Story of the New Kenyan Seed Law: 2005–2012

The road to the passage of SPVAA 2012 started with the creation of two separate and behind-the-scenes task forces constituted by the Kenyan Government's Agricultural Ministry in 2005 and 2006.[16] Their mandate was to review the 2002 SPV law (Cap 326) and develop a new one, as outlined by one of the key actors involved in both processes, namely the Seed Trade Association of Kenya. STAK was already a key actor in the Kenyan seed space, as the advocate for seed TNCs, Monsanto, Syngenta, Kenya Seed Company (KSC) and other commercial seed players operating in the now burgeoning formal seed sector[17] as well as the Secretariat for the Eastern Africa Seed Committee (EASCOM), which was represented throughout this period by STAK's CEO, Obongo Nyachae. Separate entirely from these 'shadow' task forces, an 'official' National Seed Policy Committee was formed, which included a wider group of actors. The official National Seed Policy Document of June 2010 emerged from this committee. STAK was represented on both task forces as well as the Seed Policy Committee (STAK 2007, p. 2; NSP 2010, Annex 1, p. 36).

Figure 4.1 lays out a road map that leads from the creation of the task forces to the passage of the SPV Amendment bill 2011, and subsequent enactment of SPVAA 2012. There are two strands on the map. The first is task force 1, depicted here on the left side of the map, which predominantly dealt with bringing forward a new seed policy. The seed policy was an essential precursor to the legislation itself SPVAA 2012. The second strand, which begins with the formation of the second task force (task force 2) on the right, was more specifically charged with revising Cap 326

[16] In that period 2005–2012, three different Ministers for Agriculture oversaw the events that culminated in a new seed law SPVAA 2012. They were Kipruto Arap Kirwa (2003–2007), William Ruto (2008–2010) and Sally Kosgei (2010–2013). Ruto and Sally Kosgei exchanged portfolios in April 2010, Ruto having been suspended by Prime Minister Raila Odinga on 14 February 2010, following a report by PricewaterhouseCoopers regarding a maize scam. Kosgei became Minister for Agriculture. A few months later Ruto was demoted to Minister for Higher Education, a post he held only until 19 October 2010, when he was finally relieved of his ministerial duties altogether, after a court ruled that he must stand trial over allegations of corruption, based on the new Kenyan Constitution 2010. This was separate to the ICC case against Ruto, which was dropped on 5 April 2016, due to insufficient evidence. Ruto was appointed as Deputy President of Kenya in 2013.

[17] The formal seed sector expanded at a rapid pace since liberalisation, going from just three seed companies in the 1980s, to 18 in the 1990s and post-liberalisation escalating to 78 by 2010, 90 in 2012 (AFSTA 2010).

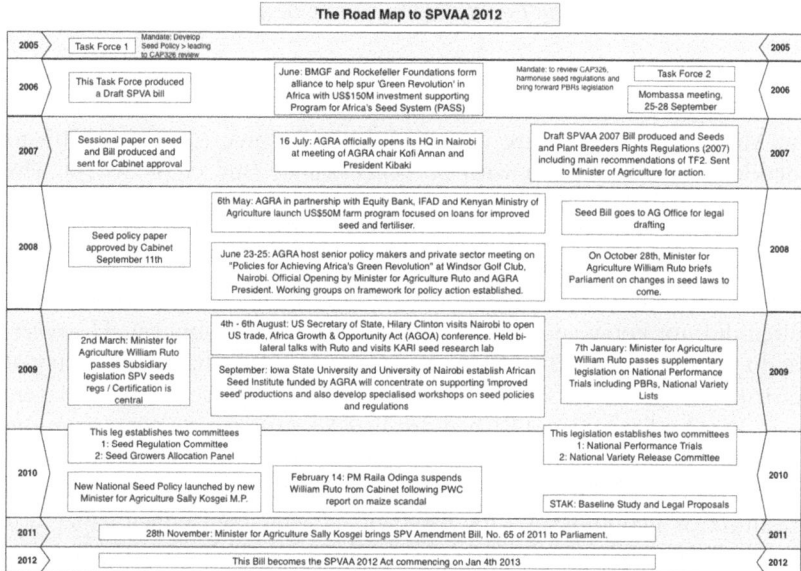

Fig. 4.1   Chronology of key events of 2005–2011 leading to SPVAA 2012

and producing a new SPVA Bill and law. Both strands dovetailed at the end with the enactment of the law itself on 4 January 2013. In the middle of the map I include other key events, actors and influencers that were significant in determining a particular emphasis on emerging seed laws. I now look more closely at the key steps on the road map.

THE ROAD TO SPVAA 2012: THE PREPARATION
FOR THE NEW SEED ACT

Between 2005 and 2011, when the Bill that would become the new Seed and Plant Varieties Act (SPVAA 2012) was published, two task forces (2005 and 2006) were established by the Ministry of Agriculture, two subsidiary pieces of legislation (2009) were passed by the Minister for Agriculture, which included the setting up of four important statutory committees (2009), and a National Seed Policy (2010) was published following the deliberations of a select Seed Policy Committee, all of which influenced the final outcome.

The Establishment of Two Task Forces to Change Seed Legislation

Task Force 1 (TF1): 2005

In the early stages of the formulation of new seed policy and legislation, the Ministry of Agriculture constituted a task force (TF1) in 2005 to "develop, in consultation with *key stake-holders* (my emphasis), a new national seed policy document to guide review of sections of the Seed and Plant Varieties Act (Cap 326)" (STAK vol 11/06, No. 2, 2007). There is no public record of the composition of TF1 and the local sources from Kenyan civil society, parliamentarians and research organisations I consulted did not know and could not locate any details about it. However, we do know from a 2007 STAK report (STAK 2007) that the Ministry of Agriculture constituted the task force and that STAK were central players on it. We also know that TF1 had produced a draft sessional paper on the national seed industry by November 2006 (ibid.). Equally, we know that a Seed Paper and Bill were forwarded to Cabinet by 2007 for approval (Institute of Economic Affairs 2008, p. 3). This seed policy paper was approved by Cabinet on 11 September 2008 (World Bank 2013, p. 85). However, it is unclear whether this paper emerged directly from TF1, or from a parallel process constituted by the Ministry of Agriculture to bring forward a National Seed Policy Document which I shall deal with later. This National Seed Policy was eventually published in September 2010 by the then Minister for Agriculture Sally Kosgei (2010–2013).

Task Force 2 (TF2): 2006

The second task force (TF2) was constituted by the Ministry of Agriculture in 2006, with a specific mandate to revise sections of the Seeds and Plant Varieties Act (Cap 326) "with a view to removing any clauses that prevent full liberalisation of the seed industry in Kenya" (STAK 2007), and with an underlying objective to bring in specific PBR legislation also. This is a key feature of UPOV 91 and is central to the agenda of the commercial/private sector, who were seeking to stimulate market liberalisation of Kenya and East Africa's seed system, through 'improved seed', including GM technologies, a position now being championed by newly established giant 'philanthrocapitalist' organisation Alliance for a Green Revolution in Africa (AGRA), which had just established its headquarters in Nairobi on 16 July 2006, and immediately became a major seed player in Kenya.

TF2 was steered by key personnel from the policy department within the Ministry of Agriculture, namely Paul Chepkwony, S.K. Angore and

Mosoti Andama, backed by other Ministry personnel from various divisions, the Ministry's legal officer, J.K. Gichuru, the research and extension liaison and the horticulture department (STAK vol. 11/06, No. 2, 2007). The newly established (2004) cross-ministerial, inter-stakeholder, donor-funded body, the Agricultural Sector Coordination Unit (ASCU), was represented by Gicheru Mucangi, while Kephis was represented by Gladys Maina. The other notable additions include Francis Ndambuki, Chairperson of Plant Breeders Association of Kenya (PBAK),[18] and Obongo Nyachae, CEO of STAK, but who was also now head of the regional seed harmonisation body and USAID-funded Eastern Africa Seed Committee (EASCOM) as well as chair of the Seed Harmonisation Committee. Both Ndambuki (PBAK) and Nyachae (STAK) from this point on become two key influencing actors in the process, and are represented on every relevant body pertaining to the formulation of SPVAA 2012. The inclusion and exclusion of key actors at this critical juncture is clearly a notable determinant in later outcomes, given their stated objectives and their precise mandate to draft the new seed law, which I will return to later. It is worth noting that one farmer organisation, the Kenyan National Federation of Agricultural Producers (KENFAP), was represented on the Seed Policy Committee, which drafted the National Seed Policy document (2010), but was excluded from TF2 (STAK 2007, p. 2), which had no farmer organisation and no civil society organisation represented at the table.

In this context, a meeting of TF2 in Mombasa from 25 to 28 September 2006 became a defining moment for changing Kenya's seed legislation. The sole purpose of this meeting was to review Kenya's Seed law (Cap 326), "including review and harmonisation of the Seeds Regulations and Plant Breeders Rights Regulations" (STAK 2007, p. 2). It was this grouping, according to STAK, that prepared the first draft of the Seeds and Plant Varieties (Amendment) Bill 2007 and the Seeds and Plant Varieties (Seeds and Plant Breeders' Rights) Regulations 2007, containing "some of the recommendations of stakeholders", and was thereafter "presented to the Minister for Agriculture for further action" (ibid.). STAK highlight key features in the Draft Seed Bill, where change was being called for which match areas identified in my last section, that is, (1) provision for

---

[18] PBAK was formed out of a conference co-hosted by STAK and UPOV in 1993, and officially registered in 1996 to lobby for the enactment of plant-breeders' rights/IPR provisions in Kenya.

accrediting private seed enterprises/individuals to undertake certain aspects of seed certification/seed-testing services, (2) enhancement of penalties for offenders, (3) adoption of some seed certification standards (UPOV compliant) agreed upon under the regional harmonisation project of ASARECA through EASCOM and (4) a provision that makes it mandatory for breeders to not only 'discover', but to also 'develop' varieties before such varieties can qualify for grant of breeders' rights. STAK conclude that "this would make Kenya's Plant Variety Protection legislation to be compliant with the International Union for Protection of New Varieties of Plants (UPOV) 1991 Convention" (STAK 2007, p. 3).

Some of the key proposals as outlined in the STAK document are identified in Table 4.1 alongside their inclusion in the final Act. The drafts that emerged out of the Mombasa meeting in particular were reiterated in the commendations by Ministers in parliament and became enshrined in the final Seed Act in 2012. This is a key finding. Critically, it has been confirmed by the CSOs and NGOs that I interviewed that they were excluded from having any part in this crucial stage of the process of formulating SPVAA 2012. Neither the draft sessional papers and bills from 2006 nor a copy of the draft SPVAA Bill of 2008 is publicly available.

At this point, there was a delay in the passage of the SPV Amendment Bill, the importance of which I shall return to later. However, the easy passage of SPVAA 2012 was greatly enhanced when the then Minister of

Table 4.1 Before and after task force meetings between ministry and private sector, 2006–2012

Task force 2: Mombasa meeting private sector and ministry draft new Seed Bill 2007	Final Seed Act SPVAA 2012
Private seed enterprises/individuals for seed certification/seed testing	Inserted at Part 5, Section 5 3B 1(a) and (b)
Enhancement of penalties for offenders	Inserted at Part 5, 3B–E and amendment of Section 29 of Cap 326 at Sections 23, 24, 25 and 33
Adoption of some seed certification standards agreed under regional harmonisation by ASARECA through EASCOM	Inserted at Section 4(b) (i) (j) (Amending Section 3 of the principal Act) (SPVAA 2012, p. 2099)
Plant breeders rights enhanced making Kenya's plant variety protection legislation compliant with UPOV 91 convention	Inserted at Sections 16 and 17

Source: Author's review of documents (ASARECA-ECAPAPA 2004; ASARECA 2007; AFSTA 2010; Minde 2004; Minde and Waithaka 2006; Waithaka and Kyotalimye 2013; World Bank 2013; STAK 2007)

Agriculture William Ruto (2008–2010) sought to pass into law two sub-sidiary or supplementary fast-track pieces of legislation which specifically dealt with aspects of a liberalised seed legislative environment which private-sector lobbyists had been pushing for (STAK 2007). These were both passed into law in 2009. In doing so, the state deepened the integra-tive process of seed certification without the rigours of a fuller parliamen-tary debate, yet managed to build judicial legitimacy while heading off potential political controversy and greatly reducing the possibility of judi-cial review in advance.

Fast Track Law 1: The Seeds and Plants Varieties (Seeds) Regulations, 2009 (Subsidiary Legislation)

This piece of subsidiary legislation[19] emerged directly from TF1. It specifi-cally established clear guidelines on the registration of seed growers (Section 8), seed certification (Section 10), seed quality and labelling (Section 15), establishing the OECD standard for cultivar purity (Section 19) and the development of a grievance procedure and penalty system through a Seeds Tribunal (Section 21), offences and penalties (Section 22) and protection from liability (Section 23), all of which enhance the proprietorial environment that shifts seed sovereignty away from farmers, especially the majority smallholder farmers, towards the corporate/com-mercial paradigm. This was in keeping with the wishes of the dominant IOs (UPOV 91, COMESA, ARIPO). It also corresponded to the agenda of the giant seed corporate actors (represented by STAK and PBAK), which were now firmly established within Kenya's seed task forces and policy committees where the seed laws were being redrafted. Meanwhile, other powerful transnational actors, such as AGRA, and powerful states, such as USAID and US Government, were simultaneously driving the same corporate seed agenda at other levels within Kenya. This piece of legislation also established two important committees, which would be instrumental in further integrating transglobal seed governance at a domestic level, preparing the ground for the final Act.

The first committee was a Seed Regulation Committee (Section 5 of the subsidiary legislation 2009) and was chaired by the Director of Agriculture and included members from Kenyan Agricultural Research Institute (KARI), KSC, Horticulture, Forestry, National Seed Quality

[19] http://www.wipo.int/edocs/lexdocs/laws/en/ke/ke011en.pdf

Control Service, one representative of Kenyan Grain Growers Cooperative Union (KGGCU) and the chief executive of the Kenyan National Farmers' Union. A sub-committee was established under the new law called the Seed Growers Allocation Panel.[20] These committees were tasked with developing seed policy, modifying seed certification standards and moderating grievance procedures amongst other things, which appears to provide some latitude to state actors to retain power and control in some key functions, which were nevertheless not sustained in the final Act SPVAA 2012.

Fast Track Law 2: The Seeds and Plant Varieties (National Performance Trials) Regulations, 2009

Another piece of supplementary legislation was passed at this time which addressed a key recommendation of the many actors in the private sector regarding plant-breeders' rights, National Performance Trials and National Variety Lists. This was passed into law and signed by Minister for Agriculture William Ruto on 7 January 2009.[21] This piece of legislation is important as it lays out the clear guidelines for the commercialisation of the seed sector, establishing the rules of the DUS[22] standard for seed, clarifying the rules regarding the national variety list which would identify the 'list of cultivars approved for release' (Section 2), and most importantly providing for the establishment of two committees: (1) a National Performance Trials Committee (Section 7) and (2) a National Variety Release Committee (Section 12). The first committee was headed and chaired by the Director of Kephis, and included a tight-knit group with one representative from STAK, one Ministry of Agriculture, two PBAK representatives and four crop specialists co-opted by the Director of Kephis. They oversaw the conduct of performance trials (Section 8) for plant varieties to be released and ensured that the DUS tests (Section 10) (IO standard) were done correctly. The second committee on National

[20] Seed Growers Allocation Panel chaired by Provincial Director of Agriculture and would include Director of Commodity research Centre, Director National Seed Quality Control Service, MD KSC, General Manager Horticulture Crops Development Authority.

[21] Special Issue: Kenya Gazette Supplement No. 12 (Legislative Supplement No. 8), 27 January 2009.

[22] DUS standard is the strict IO standard, which means that seed must be distinct, uniform and stable. The DUS standard is associated with the onslaught of commercial seed industry, necessary for monoculture practices of production.

Variety Release was headed and chaired by the Agriculture Secretary, Ministry of Agriculture and then consisted of members including the Managing Director of Kephis, the Director of research and extension at Ministry of Agriculture, CEO of STAK, Chairman of PBAK, CEO of KENFAP, one representative of an academic institution of higher learning trained in seed science and technology appointed by the Agriculture Secretary and one representative of the consumer industry appointed by the Agriculture Secretary.

The National Seed Policy 2010

The publication of the 'official' National Seed Policy marks a significant moment on the roadmap to the radical redesign of Kenya's legislative framework. NSP 2010 was informed throughout the process by a select policy committee of 'key stakeholders' directed by the Permanent Secretary of the Ministry of Agriculture, Romano Kiome. Though separate from the preceding task forces, it brought forward similar key elements, namely seed certification, plant variety protection and plant-breeders' rights, the central tenets of an enabling environment facilitating private commercial seed enterprise. We do not know if the key personnel from the Ministry of Agriculture who steered the process of TF1 were the same as those who constituted the Official Seed Policy Committee who drafted this NSP 2010 document. The   National Seed Policy (NSP) Committee was largely composed of and certainly directed by key officials from the Ministry for Agriculture. The Seed Policy Committee also included the same high-ranking officials of STAK and PBAK, namely STAK CEO, Obongo Nyachae and Francis Ndambuki of PBAK, as well as major state actors such as Kenyan Agricultural Research Institute (KARI), Kephis, the parastatal seed regulator, Kenyan National Federation of Agricultural producers (KENFAP), Kenyan Forestry Research Institute, Kenya Institute of Public Policy Research Analysis and Consumer Information Network (NSP 2010, p. iv, Annex 1, p. 36). Albeit a wider group than TF2, whose composition we do know, NSP 2010 identified key policy interventions where the government would act to counter the major 'constraints', as it saw it, in the seed sector. The Minister for Agriculture at the time, Sally Kosgei, stated:

> The National Seed Policy is the Ministry's outline of policy interventions to be pursued in order to address current constraints in the seed sub sector ...

and its implementation will need to be accompanied with a review of supportive and regulatory frameworks to govern the seed sub-sector. (Kosgei's opening remarks to NSP 2010)

The Permanent Secretary, Romano Kiome, outlined the main strategic objectives of the Ministry which informed this national seed policy as being:

to create an enabling environment for agricultural development through the review of the current policy and legal framework that would accelerate agricultural production on a sustainable basis. (Preface to NSP 2010)

He listed the laws, notably SPVA (Cap 326), but also other key legislation,[23] including subsidiary legislation that would need to be changed to provide this supportive legal framework to fulfil this strategic new direction of the nation's seed policy. Thus, NSP 2010 lays out the policy as the foundation for the seed law that is coming. To this end, we can see how provisions in the seed policy are directly linked to the seven points I identified earlier that became enshrined thereafter in law through SPVAA 2012.

NSP 2010 establishes UPOV 91 as the international frame of reference for the new Kenyan seed policy. Alongside other IOs such as OECD, WTO and TRIPS all favouring the proprietorial model, NSP 2010 steers Kenya on this narrower seed course (NSP 2010, p. 3). Regional harmonisation for trading in a liberalised seed environment is an inherent part of this programme of change within the East African Community (EAC) (NSP 2010, Section 6.4, p. 34) where Kenya is a front-runner (NSP 2010, Section 4.4 (x), p. 30). NSP 2010 heralds the key defining UPOV 91-style changes through its Declarations on PBRs and PVP/Certification. This is clear in Section 6.5, where it specifically seeks to review Cap 326 and make provision for "Payment of royalties to plant breeders" (NSP 2010, Section 6.5 (ii), p. 34), adding strict penalties against non-conformists (NSP 2010 Section 4.4, p. 29) and Section 4.3.2, which identifies seed quality control as best "effected through certification", where seed conforms to the required standards set by law and regulations (Section 4.3.2, p. 27).

Section 4.2 lays out 11 areas where government will bring all seed sectors under the net of intent to certify or at least transform them to the

<hr>

[23] Plant protection Act (Cap 324), Suppression of Noxious Weeds Act (Cap 325), Pest Control Products Act (Cap 346) and National Biosafety Act (2008).

formal sector (4.2 (viii), p. 26). This section gives sweeping powers of intervention over everything from farmers' varieties and OPVs to GM seed (NSP 4.2, pp. 25–26, 2010), all 'in collaboration now with the private sector' (NSP 2010, p. 26) in a liberalised seed industry. In so doing, the agenda is set for not only 'building capacity' for GM and other 'novel' seed technologies (Section 4.2 (v), p. 26) into seed policy objectives, but also seeks to "transform the informal seed sector into the formal one" (NSP 2010, Section 4.2 (viii), p. 26). This has implications for the unregulated/uncertified informal farming sector in a country where "80% of the rural population relies on agriculture as their primary source of livelihood" (Government of Kenya 2010, quoted in Munyi and De Jonge 2015, p. 161), and who predominantly rely on the complex, informal system of seed sovereignty throughout their communities (Interview with Robin Buruchara, CIAT 4 November 2015). This directly relates to Section 6 of SPVAA 2012, amendment of Cap 326 Section 8 concerning the agroecological value of the informal seed paradigm.

Central is the role of the nationally designated authority (NDA), namely Kephis, which Section 6.2 (ii) states will be "redefined in view of the liberalised seed sector", stating that "this will be accompanied by institutional reforms, including changes in regulations and procedures, which promote private sector participation in the seed industry". This is significant because it marks the point where the state is officially dismantling its own core functions in relation to regulation, certification and management of seed, allowing external supranational institutions and other actors to supplement and at times replace it, but also bringing both the formal and informal seed networks under the 'registration and certification' model in a way that has the potential to seriously alter the location of power and control over seed sovereignty. These policies became enshrined in SPVAA 2012.

Lastly, NSP 2010 increases the potential for conflict around PGRs, through the apparent ambiguity surrounding who will have access and share the benefits in the event of a national germplasm and conservation unit being established in Kenya (NSP 2010, Section 5.2 (xii), p. 32). In this regard, NSP 2010 specifically refers to the historically exploitative relationship with the international agricultural research institutions which thereafter is included as a whole new section in SPVAA 2012 (SPVAA 2012, Section 20). Importantly, NSP 2010 covers the seven main points, now enshrined in national seed policy, providing us with a good insight into where the SPVAA 2012 legislation will follow.

Passage Through Parliament: SPVAA 2011 Bill Becomes SPVAA 2012 Act

Following the publication of the National Seed Policy in 2010, the new Seed Bill was presented to parliament. Sponsored by Minister for Agriculture, Sally Kosgei, the first reading of SPV Amendment Bill 2011 was introduced to the legislature on 17 April 2012, following which it was ordered to be referred to the relevant departmental committees (Hansard online 17 April 2012). The second reading of SPV Amendment Bill 2011 was proposed by Kosgei and seconded by Minister for Trade, Moses Masika Wetangula, and took place on 15 August 2012. Six members of parliament spoke supporting the Bill with some concern expressed regarding the need for more emphasis on ensuring benefit sharing regarding national plant genetic resources (Luca Kipkosgei Chepkitony), rather than just breeders' rights (Millie Grace Akoth Odhiambo Mabona, Hansard online 15 August 2012). The Minister for Finance supported the Bill, calling for increased penalties to be added to act as a deterrent against those who would breach the Act. It was then agreed that SPVAA 2011 be thereafter 'committed to a Committee of the whole House' (Hansard online 15 August 2012), the following day 16 August 2012. The third and final reading led to the passage of SPV Amendment Bill, No. 65 of 2011 and took place on 28 November 2012. This was followed by the formal completion of the legislative process with the granting of Presidential Assent on 31 December 2012 and commencement thereafter by legal notice on 4 January 2013 (Kenya Gazette Supplement No. 217 [Acts No. 53]).

I now lay out in greater detail the key actors involved in the decision-making process and whose role was instrumental in determining particular seed outcomes at the crucial time when the law was being formulated and passed.

The Key Actors Behind the New Kenyan Seed Law (SPVAA 2012)

This section examines the role of the Kenyan Government as the key state actor in the passage of the bill as well as external actors, including other domestic state actors, IOs and TNAs.

Whilst it seems that the text of SPVAA 2012 was already largely completed as far back as 2007/2008, the delay in its passage signals that some areas needed to be addressed and that other key stakeholders needed to be convinced before the passage of the new law in 2013 through the Kenyan Parliament.

From the outset, UPOV 91 was a key driver, dominating the push for the changes realised in the Kenyan Seed law SPVAA 2012. Kenya became a member of UPOV 78 on 13 May 1999 and WTO in 1995 (WTO under TRIPS (Article 27.3 (b)) requires IP PVP architecture). SPVAA 2012 marks the domestication of UPOV rules. It amended its domestic legislation to bring Kenya in line with UPOV 91, which is characterised by its pro-plant-breeder IP proprietorial framework, and the DUS test for seeds. The latest revision of these rules in 1991 further strengthened PBRs, granting monopoly rights over 'discovered' varieties and the production, marketing, export and import thereof, and bringing prohibitive restrictions on farmer practices of seed sovereignty. On 28 November 2011, Sally Kosgei, echoing her predecessor William Ruto's line on the 'need for harmonisation', laid out the 'memorandum of objects and reasons' for SPVAA 2012, stating that the reforms are necessary so that the 'seeds subsector is harmonised with the current policy environment in order to promote a modern and competitive seed industry' (Kosgei in kenyalaw.org 28 November 2011). Clauses 15, 16, 17, 18 and 26 specifically refer to the need to comply with UPOV 91. The realisation of UPOV 91 rules in SPVAA 2012 set the framework and the backdrop for all the other changes to come, because it encapsulated stricter PBRs, PVP, variety release, duration of testing periods for key crops, registration, certification standards and everything that favoured commercial seed enterprise over farmer/indigenous seed systems. It marked a deep shift in seed sovereignty. Domesticating these changes would require significant adjustment to the present regulatory regime governing seed within the state itself. This put Kephis, as the national regulator, centre stage.

The changes sought potentially pitched the Government of Kenya against other state actors, namely parastatals such as Kephis, the Kenyan Research Institute (KARI), the Kenya Seed Company (KSC)[24] and any other body which could be and certainly in the case of Kephis was regarded as a 'bottleneck' to private seed sector development (USDA 2008; ASARECA 2007; Minde 2004; Minde and Waithaka 2006; ASARECA-ECAPAPA 2004, p. 1; STAK 2007, p. 3).

---

[24] KSC has had a monopoly on production and distribution of new varieties released by state research organisation KARI, which was mandated to develop new crop varieties.

We know that the US and TNC seed actors, represented by STAK/
EASCOM and PBAK, were particularly exercised about the role of
Kephis,[25] seeing its monopoly role as an impediment to private-sector
innovation and growth, and US Government documents at that time
insisted that it "has usurped the developmental role in the industry"
(USDA 2008). We also know that there were vast profits at stake for these
stakeholders.[26]

As highlighted above, STAK was formed in 1982[27] to represent the
interests of the seed TNCs, Monsanto[28] (the world's largest seed company), Syngenta, Bayer, Pioneer Hi-Bred, Pannar Seed, Kenya Seed
Company (KSC) and other registered seed enterprises, advocating for
UPOV 91 and the development of the formal seed sector.[29] Closely linked
to STAK, PBAK was formed out of a conference co-hosted by STAK and
UPOV in 1993, specifically to lobby for the enactment of plant-breeders'
rights provisions in SPVA in Kenya. "PBAK gave lobbying for enacting
PBR provisions of SPVA a collective voice, while also channelling expertise
to shape the architecture of law" (Rangnekar 2014, quoting Louwaars
et al. 2005, p. 3).

STAK CEO and EASCOM chair, Obongo Nyachae, was a central actor
at every forum pushing for key changes in the role of Kephis but also on
all the other major points featured in NSP 2010, for which they sought

[25] https://msu.edu/course/aec/841/Hort-FullReport16Feb2004_web3.pdf

[26] GRAIN report recognises "huge potential for seed exports from the U.S. as the formal
seed sector expands and the demand for numerous seeds grows" (USDA 2008). GRAIN
point out that African Seed Trade Association (AFSTA), which was set up by the American
Seed Trade Association (ASTA) "to act a s a local lobby for the transnational seed industry"
was mandated to "promote regional integration and harmonisation of seed policies and regulations supportive of US seed trade", with an explicit target of securing a 5% increase in US
seed exports to the region in the first 5 years (GRAIN 2005, p. 31).

[27] STAK was formed in December 1982 under the Societies Act Cap. 108 of the Laws of
Kenya.

[28] "For the financial year ended August 2014, Monsanto's revenue topped US$15.8 bil-
lion, with total profits exceeding US$ 2.7 billion. R&D expenses for the year were US$1.7
billion" (Monsanto quoted in ACB 2015, p. 36).

[29] STAK is a member of the International Seed Federation (ISF), is linked to the African
Seed Trade Association (AFSTA), an African apex body representing the private sector in the
seed industry, which in turn has strong links to the American Seed Trade Association, and
expands through the African Regional Intellectual Property Organisation (ARIPO), the
Organisation Africaine de la Propriété Intellectuelle (OAPI), the regional intellectual property organisation for mainly Francophone African states and the Southern African
Development Community (SADC) to the rest of Africa.

resolution in SPVAA 2012. To this end, Waithaka and Kyotalimye (2013) assert that EASCOM in particular:

> supported the national stakeholder review processes, provided technical backstopping and legal support to the drafting process, and facilitated parliamentary retreats and media releases. (Waithaka and Kyotalimye 2013)[30]

They were particularly active at a regional level also through COMESA, ARIPO and similar bodies, constantly calling for UPOV 91 compliance through domestic legislation and regional harmonisation which would be legally binding. This push for regional harmonisation was uppermost in Sally Kosgei's agenda when presenting the SPVAA to parliament in 2011, stating that members within Kenya's trading bloc had acceded to UPOV 1991, but "we have not and that is what we seek to do now" (Kosgei in Hansard online 15 August 2012).[31] ARIPO is the regional IP PBRs driver for Anglophone countries, including Kenya. Established under the Lusaka Agreement in 1976, and anchored in UPOV 91, ARIPO's Council of Ministers approved a proposal for ARIPO to develop a policy and legal framework in November 2009, which would form the basis for the development of the ARIPO Protocol[32] on the Protection of New Varieties of Plants (the PVP Protocol), and develop harmonised procedures for registering PBRs for all ARIPO members.

This push for harmonisation with UPOV 91 and regional accession through ARIPO, COMESA, ASARECA and EASCOM was unrelenting during this period 2005–2013. Kenya, as a selected 'pilot' country for a 'Seed Initiative' and a 'Policy Change Cycle' since 1999 (ECAPAPA–IFPRI Brief 2006), was regarded at these fora as a test case for wider accession to a stricter seed regime within Africa (Odame and Muange 2011; ASARECA 2007; Minde 2004; Minde and Waithaka 2006; ASARECA-ECAPAPA 2004, p. 1; STAK 2007). This project was greatly facilitated by the construction of the private task forces in 2005 and 2006.

A strong government actor to direct domestic changes at the state level was needed to bring the other stakeholders on board to ensure accession

---

[30] http://knowledge.cta.int/Dossiers/S-T-Issues/Seed-systems/Feature-articles/Harmonising-seed-policy-in-Eastern-and-Central-Africa-lessons-from-a-public-private-partnership-model

[31] In fact, only Tanzania within the ARIPO countries has fully acceded to UPOV 91.

[32] This officially became the Arusha Protocol in July 2015, though remains contested whether it can have direct legal application (Munyi and De Jonge 2015).

to the demands of an intensifying private-sector lobbying. This actor was William Ruto, who was appointed as Minister for Agriculture on 17 April 2008. Ruto was already a skilled politician and well versed in his new brief having studied botany and graduated from Nairobi University with a Master's degree in Environmental Science. Through him initially, and followed by his successor, Sally Kosgei, the Government of Kenya asserted its role as the primary state actor. For the next few years Ruto was the main champion of the state transition for "innovative, commercially-oriented and modern farming as a business" (Minister for Agriculture William Ruto MP addresses Kenyan Parliament in Hansard online 28 October 2008). Ruto immediately signalled state intent to "harmonise legislation and rationalise the seed sector … to help us to manage the sector better and reduce the areas of conflict" (ibid.). Harmonisation meant UPOV 91 rules.

Both Ruto and Kosgei, as evidenced from parliamentary debates (Ruto in Hansard online 28 August 2008, Kosgei Hansard online 15 August 2012 and interview 6 November 2015), viewed changing the seed law as central to the fulfilment of the national project of economic liberalisation,[33] which sets Kenya, and indeed East Africa on track for a Green Revolution. 'Green Revolution' means corporate seeds, not farmer's seeds, and was a direct statement of intent to shift seed sovereignty away from farmers in any policy and legislative reform. We can see this when, within weeks of his appointment speaking at an AGRA Convention on "Policies for an African Green Revolution" on 28 June 2008, Ruto took his host's line stating:

> The current world food crisis has provided a wakeup call for the policy makers to reorient their planning process to provide viable and sustainable solutions … (for) a green revolution which will dramatically increase Agricultural Productivity and lift the bulk of our population out of poverty. (Ruto, AGRA Press Release, 28 June 2008)

---

[33] Kenyan State is deeply committed to this paradigm at many levels. Through its Vision 2030 blueprint for Kenya's Development, in its Strategy for Revitalising Agriculture (SRA) 2004–2014 and its 2010–2020 Agriculture Sector Development Strategy. Ruto also refers to Kenya's commitment under the Maputo 2003 Declaration signed by African heads of state to increase agriculture budget to at least 10% of national budget. They are also committed to this paradigm through COMESA, ARIPO, Alliance for Commodity Trade in East and Southern Africa (ACTESA), Comprehensive Africa Agriculture Development Programme (CAADP), ASARECA at a regional level and through these initiatives and treaties they are in turn aligned with their key resource partners, namely, USAID, AGRA, EC, World Bank, Department for International Development (DFID) and others, who all share the same paradigmatic approach to seed.

AGRA, which had only just established their headquarters in Kenya in July 2007, was already investing hundreds of millions of dollars at every level in the Kenyan seed sector. This included pushing for regulatory reform, promoting new 'improved seed' and GM seed solutions and funding 80 new seed companies through their Programme for Africa's Seed System (PASS) (AGRA 2014 quoted in ACB 2015, p. 18). Meanwhile, the Gates Foundation bought $23 million worth of Monsanto shares in 2009, and was now directing 30% of their agricultural development funds to developing genetically engineered seed solutions for sub-Saharan Africa.

By October 2008 and just one month after the Cabinet received the papers from TF1 on seed policy in 2008, and while the TF2 Seed Bill was in the Attorney General's Office, Ruto, who was also the Chairman of the African Council of Ministers of Agriculture at this time, briefed parliament on the legislative reforms his Ministry had been working on for the past three years (without reference to the task forces, or their composition) (Hansard online 28 October 2008). Ruto outlined his intention "to legally empower *our stakeholders* (my emphasis) through improvement of the current agricultural legal and regulatory framework ... to ensure that all players adhere to set standards and regulations" explaining that "the process of policy formulation and Bill drafting takes time", because his stated purpose was to "review, amend or repeal numerous Acts"[34] (Ruto in Hansard online 28 October 2008).

He singled out Kephis in this speech to parliament, outlining his proposal to change the regulatory function of what his successor Sally Kosgei (2010–2013) describes as Kenya's 'major seed stakeholder' (Interview with Sally Kosgei, 6 November 2015):

> I will bring to this house legislation to amend the Act under which the Kenyan Plant and Health Inspectorate Services (Kephis) operate so that we can add additional responsibility to the organisation. (Ruto parliamentary speech in Hansard online 28 October 2008)

In fact, core functions of the organisation were being opened up to private-sector interests under SPVAA 2012, Section 3B (b). This dismantling of a core function of the Kenyan State, shifting seed sovereignty from a publicly accountable institution of the state and opening up key aspects

---

[34] One-hundred and thirty pieces of legislation would be reduced to 7 according to Ruto (Hansard 28 October 2008).

to the private realm, was central to Ruto's agenda when he addressed the Kenyan Parliament (Hansard online 28 October 2008). In addition, Ruto's enactment of two vital pieces of subsidiary legislation in 2009 paved the way for the deeper changes he sought in altering the core functions of Kephis. According to Waithaka and Kyotalimye (2013), the passage of both of these pieces of legislation, namely 'The national performance trials regulations' (Kenya NPT regulations 2009) and 'the Seeds and Plants Varieties (seeds) regulations 2009', were lauded by the USAID-funded body EASCOM as significant legislative achievements. They were particularly pleased with the fast-track legislation brought in by Ruto, which 'set the pace for accession to UPOV 1991' (ibid.), which directly links bodies such as STAK/EASCOM and PBAK with the juridical changes inside Kenya from the inception, and a clear indicator of how the intervention intensified under Ruto.

Coinciding with the passage of these key fast-track regulations, Hilary Clinton made a more than symbolic visit to Kenya in August 2009. As US Secretary of State, she was joined by head of USDA, Tom Vilsack, and held bilateral talks with William Ruto and opened the Africa Growth and Opportunity Act (AGOA) conference, followed by a meeting to KARI, where Tom Vilsack called for 'transformational change'. Clinton emphasised the strong 40-year link between the USA and KARI, stating that USAID had provided more than $4 million to KARI since 2003 for 'agricultural technology development and transfer'. She went on to state that agriculture in Africa had been held back by "slow adaptation to technology, as well as a lack of investment from the private sector" (*Science Africa*, Vol. 7, August–September 2009, available at https://scienceafrica.co.ke/wp-content/uploads/2018/02/siav7.pdf. Accessed 26 December 2018).[35]

Sally Kosgei, who took over from Ruto as Agriculture Minister in 2010, explains her thinking in bringing forward the SPVA Bill. She cites a trip to Iowa in place of the then President Moi, following an invitation from the World Food Programme, as an important moment.

---

[35] This was echoed in 2009 US President Barack Obama highlighted developmental 'game changers' in Africa in his US National Security Strategy speech, citing the need to pursue issues such as agricultural productivity which are 'not adequately addressed at a bilateral level', emphasising the need to pursue the potential for weather-resistant seed varieties and green energy technologies as an example https://www.whitehouse.gov/sites/default/files/rss_viewer/national_security_strategy.pdf (Obama US National Security Strategy 2009, p. 34).

They have a huge seed industry in Iowa. Technology has worked very well for the United States. It is the foremost nation on science and technology and didn't get where it is today without it. I myself was educated in the US, so I wanted to have a law certified so that down the line people cannot say that we didn't know what was coming in. (Interview with Kosgei 6 November 2015)

Certification was central. It would ensure plant-breeders' rights and plant variety protection fulfilling UPOV rules and lay behind the rationale of changing the regulatory functions of Kephis. Kosgei stated that:

The major stakeholders involved in the SPVAA process were Kenyan Agricultural Research Institute (KARI), Kephis and the private seed companies who needed to be certified. (Interview with Sally Kosgei 6 November 2015)

Kosgei explained that as Minister for Agriculture, her:

thinking behind the Act was based on the need to protect our own seed, protect our farmers and to certify seed, especially maize … our commercial crops and we also were protecting our borders. (Interview with Sally Kosgei 6 November 2015)

She went on to explain that she had previously been the Minister for Science and Technology and had passed the Biosafety Law, which ensured that things would be certified and regulated, though she claims it was misunderstood. So when she became the Minister for Agriculture, she ordered the research institutions to work on building this protection from the outside and from the inside. For Kosgei, it is implied in the 2012 Seed and Plant Varieties (Amendment) Bill that changes were coming in from the outside, and that her main motivation was 'to protect Kenya's seeds' (Interview with Sally Kosgei 6 November 2015). Certifying seed for commercial purposes, for standardising regional seed trading and harmonising national legislation in keeping with UPOV 91 and COMESA, the strictest international and regional standards was axiomatic to any new seed law, and Sally Kosgei was the main actor who secured those changes.

Behind the frontline political process lay the main research institutions which the Kenyan State looked to at that time. One such body was the International Centre for Tropical Agriculture/Consultative Group on International Agricultural Research (CIAT/CGIAR), which "has a big influence on policy" according to CIAT's Africa Head, Robin Buruchara (Interview with Robin Buruchara Africa head of CIAT 4 November 2015). They "provide the evidence needed to make adjustments at the bureaucratic level of laws or to justify private sector involvement in some cases" (ibid.). This is extremely important in the wider context of seed sovereignty, especially given the increasing philanthrocapitalist funding and influence on CIAT/CGIAR itself.

Speaking specifically about the formulation and passage of SPVAA 2012, Buruchara stated that "the process of developing an Act is very bureaucratic", but he stated CIAT/CGIAR see stimulating free markets and private-sector engagement at every level of seed production from breeder seed to basic or foundation seed for multiplication purposes as central to developing a vibrant seed system regionally. To this end, CIAT is engaged with numerous central players, USAID, Syngenta and AGRA.[36] CIAT's role is "to generate the evidence needed for example to show that the fear was not justified in allowing private sector seed to be involved a little earlier in the seed process for example", so even if the law does not change, "some of those things are already changed" (ibid.). CIAT at that time urged the need to stimulate interest in key crops at different stages of seed/crop dissemination. Though Buruchara was unimpressed by the bureaucratic nature of laws and regulations, he saw the SPVAA 2012 law as designed "to protect farmers from crooks", echoing the view put forward by Kosgei and other parliamentarians. CIAT/CGIAR clearly played an influential role in providing evidence to convince other actors, particularly Kephis, of the need for change in the key areas which the private sector in particular was pushing for, and which bodies such as ASARECA, COMESA, USAID, AGRA and others were pushing to ensure rapid domestication. CIAT was integrally involved with this drive, through its project work but also behind the scenes in the various fora where legislative changes such as SPVAA were conceived. Buruchara explained that CIAT has ongoing projects with many of these actors. These include collaboration with transnational seed corporation, Syngenta, on a pan-African

---

[36] CIAT was 'overhauled' in 2000. This led to the creation of a CGIAR fund and saw its 15 global research programmes rehoused in the World Bank in Washington DC and CGIAR entered into a number of public-private partnerships (PPPs) with seed MNCs Monsanto, DuPont, Pioneer and Syngenta since, notably in Kenya (ACB 2015, p. 10). ACB highlight how the organisation is hugely funded now by AGRA to the tune of US $720 million from 2003 to 2014, while the Bill and Melinda Gates Foundation hold a place for BMGF on the CGIAR Council (ibid.).

Bean Research Alliance,[37] and a joint project with USAID-building models for seed dissemination, while they work with AGRA on the 'Africa Rising Programme' in the region. The AGRA/CIAT programme focuses on "facilitating the development of the small private sector seed companies, to fill the niches in different areas enabling seed to be accessed more easily". CIAT also work at a regional level with COMESA on seed matters according to Buruchara.

By August 2012, when the Bill was formally introduced for its second reading in parliament, Kosgei stated that

> The challenges facing the seed industry in Kenya include inadequate coverage of some varieties and plants in the seed certification process, lack of comprehensive breeders' rights and lack of definition of the competent authority on matters of seed certification, among others. (Kosgei in Hansard online 15 August 2012)

She went on to note that Cap 326 did not provide for Kephis as the competent authority, stating that "this needed to be and was being resolved in the new seed law" (ibid.).

In this parliamentary address, it becomes clear that any unease regarding Kephis' changing role had been dealt with. Kosgei announced the "authorisation and registration of private seed inspectors and seed-testing services to supplement the services offered by Kephis, thereby encouraging self-regulation in the industry" (ibid.). Interestingly, the Minister for Trade who seconded her remarks pointedly states that "Kephis should thereafter 'quickly conform to the standardisation that is ongoing within the context of the East Africa Community'" (Minister Moses Masika Wetangula, Hansard online, 15 August 2012). The government consolidated its position, ceding a key area of seed sovereignty by bringing a separate Bill on Kephis before parliament the same day,[38] which firmly

---

[37] http://ciatblogs.cgiar.org/support/syngenta-foundation-stronger-bean-seed-systems/ [accessed online /6/2017].

[38] Kenya Gazette Supplement, No. 218 (Acts No. 54). The President would appoint the non-executive chairperson of the new 'Service', which would now be responsible for "implementing PVP in Kenya, and administering PBRs (Part 11, Section 5 (f), Kenya Gazette Supplement, No. 218 Acts No. 54), and undertaking plant variety testing and description, seed certification (Part 11, Section 5 (g)) (ibid.), and to implement and enforce national biosafety regulations on the introduction and use of genetically or living modified species of plants, insects and micro-organisms, plant products and other related species" (Section 5 (j)) (ibid.).

established its modus operandi and its structure to conform to a new commercially driven privatised seed culture, dictated by SPVAA 2012.

Regarding other state actors, the role of the Kenya Seed Company and Kenyan Agricultural Research Institute is significant in relation to PBRs, another central ingredient of seed sovereignty and a signal of critical change in the Kenyan context at that time. Both public bodies had enjoyed a virtual national monopoly. KARI was the mandated research institution responsible for developing new crop varieties, which KSC then produced and distributed. Rangnekar (2014) points out that earlier liberalisation of Kenya's seed industry in the mid-1990s had ensured that public and private breeders had a shared position on the issue of plant-breeders' rights (PBRs). Following liberalisation of the seed sector in 1995, Rangenkar also points out that this monopoly was terminated. An influential Deloitte and Touche Consulting Report at the time signalled the potential for vast royalty revenue streams which liberalisation of the seed sector and opening up to licencing new varieties would bring (ibid., p. 374). This was an essential ingredient in moving SPVAA 2012 forward, with agreement amongst such essential domestic actors. This would explain why seemingly important domestic actors, KSC and KARI, were not on the task forces. STAK, PBAK and the government were adequately representing their wishes for stricter PBRs, as 'the surge in applications flooded in', and royalty payments flowed out (ibid.). This was all ambiguously supported by Kenya's main public seed breeder, and greatly favoured by private actors eager to apply for PBRs, and who stood to make vast amounts of money. "The greatest beneficiaries of PBRs in Kenya are external breeders", as ASARECA confirm in their report at the time (Waithaka et al. 2011, p. 14), and echoed by STAK CEO during his address at the Open Forum on Agricultural Biotechnology in Africa (OFAB) on 28 June 2007.[39] Another key area of domestic seed sovereignty was thus opened up to external actors, whose market power would allow them to benefit considerably from Kenya's new PBRs arrangements under SPVAA 2012.

However, former Assistant Minister for Agriculture (2008–2013), and Agricultural Committee member, Japhet Mbiuki who served under both Ministers, states that:

[39] http://www.slideserve.com/gage-bauer/presentation-made-during-8-th-ofab-meeting-in-nairobi-28-th-june-2007 [accessed online 8 June 2017].

The Ministry of Agriculture was the main driver throughout the SPV Amendment process, and the State is the major player because at the end of the day, the Government is 100% in charge. (Interview with Japhet Mbiuki former Junior Minister for Agriculture 6 November 2015)

Mbiuki points to the fact that Kenya Seed Company, the state company, is the largest seed producer, and therefore an important player (ibid.). He does not refer to their unique and ambiguous position, where they are now reaping vast rewards from private-sector, PBR-inspired rules under SPVAA 2012. When asked about the role of the private-sector actors, Mbiuki states:

> Of course we brought in the private players, whose views cannot be ignored to ensure that they are also properly captured, such as KSC, Kephis, Monsanto, Similo, Kenyan Bureau of Standards, Maize Farmers Association, consumer organisation COFEK and farmers. ... Actually we are looking at what is good for our country, without undue influence from multinationals. The multinationals are going to come and lobby but basically we are looking at something which has a national view and something which will be able to ensure that there is food security in the country. Our bottom line is food security, not to play into the hands of multinationals. (ibid.)

However, his parliamentary colleague on the Agricultural Committee (AC),[40] Dr. Victor Munyaka, highlights that Monsanto was a major stakeholder, already controlling 20% of the hybrid seed sector in Kenya (Interview with Munyaka 6 November 2015). This is a reference to the already liberalised nature of Kenya's formal seed sector, which has enjoyed an open door to TNC actors like Monsanto for many years, earning it the title of "'poster child' for Africa's new Green Revolution" in IDS Bulletin's special edition in 2011 (Odame and Muange 2011, p. 78). Munyaka asserts that "Monsanto is a world leader that is supplying to agrodealers in a network all over the country", stating that the Agricultural Committee met with Monsanto in Washington in 2013 (Interview with Munyaka 6 November 2015). Munyaka adds that Syngenta, Bayer and BASF are also important players, whilst pointing out that the traditional or orphaned crops are missing out in most laws, and need to be exempted from taxes so that "we can revive and multiply them because they are so nutritious for the people" (ibid.).

---

[40] The 29 strong Agricultural Committee is an important drafting committee for agricultural legislation such as the seed bill.

However, farmer seeds and traditional crops were not part of the SPVAA 2012 agenda, as confirmed by Sally Kosgei in her interview when she clearly states that "we were not concerned about traditional crops" (Kosgei interview 6 November 2015). This was also confirmed by MP Bonny Khalwale who states that the 'traditional farmers' were not consulted at all:

> In fact the only time they get to be involved is when they are using the product—they are told this is a new seed that is recommended—buy it. Nobody consults them. They are not even in the legislation. (Interview with Dr Bonny Khalwale at the Kenyan Parliament 6 November 2015).

So despite Kosgei's assertions of wanting to 'protect our farmers', she continued Ruto's example of not including them in the decision-making process which led to the new seed law. Odame and Muange (2011) assert that the first draft of the SPV Bill the Ministry of Agriculture sought to have the informal seed system included in the new seed law, in recognition of its importance in seed supply. However, during subsequent deliberations, many seed companies were reluctant to let the informal system be included in a revised version of the law:

> They feared loss of market share due to the anticipated competition from a 'better organised' informal sector in the market, and intensive lobbying by the companies contributed to slowing down the process of enacting the law. (Odame and Muange 2011, p. 85)

Yet there is no evidence that this was in fact the case or the reason for the deliberative process over a few years. Neither the Minister, nor any number of MPs, nor the record of STAK, EASCOM or ASARECA from that period suggests that Kenya's SPV Amendment Bill or Act was ever going to include or be designed with 'farmer seeds' in mind. In fact, STAK records from as early as the first draft in 2007 express satisfaction with the content of the new draft law, suggesting that the informal system and farmer seeds were not included. This is at the centre of the contest over seed sovereignty at this time.

The African Biodiversity Network (ABN) sees SPVAA 2012 as being "aligned to the injurious requirements of UPOV 91" and represents laws that are "taking away the responsibility for African peoples to feed themselves, and giving it to multinational companies, whose key interest is to make profits" (ABN 2014, pp. 8–9). Yet, the former head of the Human Rights Commission, Senator Hassan Omar Hassan, asserts that the Kenyan State is strong, quoting the line that the Kenyans alongside the Nigerians and the South Africans have a greater sense of institutional memory, a better sense of activism, both in the social media and the political medium, and is difficult to suppress (Interview with Hassan 6 November 2015). Yet, neither civil society nor the majority farming population was consulted at crucial decision-making stages of the process of drafting this new seed law. This has profound implications for the seed practices and agricultural biodiversity of the majority farming population of the entire country, thus making Hassan's assertion meaningless in the context of the fate of Kenya's seed sovereignty, and thereby its control over its nation's food.

The Berne Declaration (2014), which carried out a human rights impact assessment of the impact of UPOV 91-based legislation, concluded that "no association (including farmers' associations) reported having been consulted or contributing to the process leading to the enactment" (Berne Declaration 2014, p. 37). They state that their research team "could find no evidence that the government mandated any assessments of the likely impacts of UPOV 91 type legislation" (ibid.). They found that "the Kenyan Farmers' Association (KENFAP) complained of not having been made aware of, involved in or consulted in these processes" (ibid.). This corresponds to a recent report of the World Bank (2013), which reflects similar unease amongst some participant/stakeholders, who feel that their views are not always reflected in final policy and legal documents. They state that some interviewees noted "that policies could be implemented in ways that differed from the consensus reached by stakeholders" (World Bank 2013, pp. 75–76). Yet the World Bank themselves were deeply involved with the programme to change the seed regulations through their sub-Saharan Africa seed Initiative.

This effectively allowed the state, led by government to act as a unitary actor, delivering a new seed law, which betrayed no signs of any discontent from within the state itself, yet delivered the changes that powerful external actors had been seeking, which has grave implications for the principles and practices of human rights of the most vulnerable groups in sub-Saharan Africa and their seed sovereignty.

Conclusions from the Kenyan Case

Testing the approaches to globalisation against the process and outcome of Kenya's new seed law SPVAA 2012 reveals a hyperglobalised seed law. Instead of adopting a 'sui generis' law, which was within its privilege under UPOV 78 rules and which it was party to, the Kenyan state chose to adhere to the strictest international standard based on the most globalised rule system, namely IO UPOV 91. Kenya thus ceded proprietorial rights over seed to corporate and commercial interests. It specifically enshrined the much sought-after plant-breeders' rights and plant variety protection, the two central hallmarks of proprietorial globalised seed legislation, to those corporate/commercial interests.

This 'hollowing out' of the state's role in the public seed policy programme was facilitated by a state willing to relinquish its sovereign control over its seed to external actors. In fact, it was championed by successive agriculture ministers, Ruto and Kosgei, in keeping with the vision of President Kibaki for a revised agricultural programme. The spoils were enjoyed predominantly by transnational actors, specifically transnational corporations, whose stated goals, as revealed in their own private documentation, were all achieved through the process and enactment of SPVAA 2012. Specifically, the 'shadow task forces' established by the Kenyan Agriculture Minister in 2005 and 2006 with a special mandate to 'revise the old seed law, removing any clauses that prevent full liberalisation of the seed industry in Kenya' were central to the construction of a hyperglobalised law.

These task forces became key instrumental channels allowing important transnational actors, representing seed TNCs (notably Monsanto and Syngenta through STAK and PBAK) to present their case at all of the main decision-making meetings throughout the drafting process, right through to the judicial stages. These powerful actors, STAK and PBAK, and their key representatives were the only other actors, apart from government representatives (even the agricultural Ministers changed three times), who sat on every important committee and task force drafting the new seed law and policy throughout this time. Their globalising agenda for seed changes were all met, namely, PBRs, PVP, certified seed, UPOV 91 and key definitional changes which favour private commercial interest over the public, informal farmer seed system, which was decidedly excluded.

These task forces were followed by the passage of two important pieces of subsidiary seed legislation in 2009 and a new seed policy in 2010, which prepared the ground and copper-fastened key aspects of the final highly globalised seed law. The changes they brought were eagerly sought and lauded by other key state actors such as USAID and USDA, who had a clear commercial agenda and was closely aligned with the powerful TNC actors, notably US seed giants Monsanto and Syngenta, as well as the US philanthrocapitalist body and key TNA, AGRA.

The same external actors, notably STAK and PBAK, were pushing a similar agenda at the regional level, seeking transboundary ease of passage and seed harmonisation through COMESA, ECAPAPA and ASARECA, amongst others, all backed by the same powerful external actors, who relentlessly pushed for and consistently financed these background meetings, organisations and associations. They sought seed legislative harmonisation across the region and Kenya was a priority pilot country, important in promoting their agenda in the region and on the continent.

The domestic state colluded throughout the process with its own retreat from sovereign control over seed. It did this without consultation with the 80% of the smallholder farming population who rely on the informal seed network, thus also denying them sovereign control over their seed systems. This exclusion of the informal 'farmer-managed' seed system by government in the revised law is significant. Their legislative bias in favour of the formal (corporate seed) sector is at the root of the contest between farmers' rights enshrining practices of seed sovereignty and a proprietorial plant-breeders' rights paradigm, epitomised by rules of certification and penalties for infringement, central hallmarks of globalising IP law. This is now domesticated in SPVAA 2012, signalling a major shift in seed sovereignty away from the commons/public arena to supranational agencies and transnational actors, outside of publicly accountable systems. In doing so, the Kenyan State chose to ignore the principles of farmers' rights enshrined in other global rules such as the CBD and the ITPGRFA, in favour of UPOV 91. It therefore pitched farmers and their seeds against corporations and their corporate seeds (hybrid and GM), and the necessary regulatory infrastructure which goes with the latter.

The passage of SPVAA 2012 marked the moment when Kenyan seed sovereignty shifted to powerful external actors, in keeping with hyperglobalist thinking. These 'shadow sovereigns' now exercise rights over formerly public seed systems and determine seed policy futures and seed practice. The Kenyan State chose to follow the dictate of leading IO, UPOV 91, key TNAs, namely the giant seed companies such as Monsanto and Syngenta fronted by STAK and PBAK and adhered to the wishes of its key geopolitical ally, the USA in determining the most important change in its seed law in over a decade. It underlines how through a rigorous and lengthy process of 'socialisation around IPRs' amongst key actors public and private, consensus was built well before parliamentary debates were held.

Hyperglobalism clearly best explains the role and influence of actors in changing the seed law in 2012. The Kenyan State was complicit in its own withdrawal from the public seed programme, but not in a manner that sceptics could assert demonstrated an assertion of state control. Neither did it represent the contradictory and ambiguous way that transformationalists assert is the modus of the globalising process. It allowed key transnational seed corporations, notably Monsanto and Syngenta, major roles on 'shadow task forces' which determined critical shifts in seed sovereignty, including the dismantling and reconfiguring of the state regulator, causing a hollowing out of a hitherto core sovereign state role and function, all central to hyperglobalist accounts. Those actors assumed sovereignty over Kenyan seed through a lengthy process over years leading to a hyperglobalised seed law, SPVAA 2012.

REFERENCES

ACB. 2015. *The expansion of the commercial seed sector in sub-Saharan Africa: major players, key issues and trends* [Online]. Available from: www.acbio.org.za [Accessed 3rd February 2016].

African Biodiversity Network (ABN). 2014. ABN News. *The new Scramble for Africa.* Seeds Edition. April–June 2014.

AFSTA. 2010. *Baseline study/survey report on the seed sector in Kenya.* Nairobi, Kenya: Seed Trade Association of Kenya.

ASARECA. 2007. *ASARECA's strategic plan 2007–2016.* ASARECA.

ASARECA-ECAPAPA. 2004. Proceedings of the workshop on promoting harmonized policies and procedures for quality seed production and trade in Eastern Africa. *IN: ASARECA-ECAPAPA. 28–29 June 2004, Novotel Mount Meru, Arusha, Tanzania.*

Berne Declaration. 2014. *Owning seeds, accessing food, a human rights impact assessment of UPOV 1991. Based on case studies in Kenya, Peru and the Philippines.* Zurich, Switzerland: Berne Declaration.

Dutfield, G. 2011. Food, biological diversity and intellectual property: the role of the International Union for the Protection of New Varieties of Plants (UPOV). *Intellectual Property Issue,* Paper No. 9. Quaker United Nations Office.

ECAPAPA-IFPRI Brief. 2006. Based on Minde I. 2004. *Harmonizing seed policies and regulations in Eastern Africa: experiences and lessons learned.* Entebbe, Uganda: Eastern and Central African Programme for Agricultural Policy Analysis. https://www.ifpri.org/themes/ecapapa.htm (accessed online 2012).

GRAIN. 2005. *Africa's seed laws: red carpet for the corporations* [Online]. Available from: https://www.grain.org/article/entries/540-africa-s-seeds-laws-red-carpet-for-corporations [Accessed 11th November 2014].

IDS Bulletin. 2011. *The politics of seed in Africa's green revolution.* Oxford: Wiley and Blackwell.

Institute of Economic Affairs. 2008. *The Budget Focus. A Publication of the IEA Budget Information Programme.* Issue No. 21. May 2008.

Karembu, M., F. Nguthi and H. Ismail. 2009. *Biotech Crops in Africa: The Final Frontier.* Nairobi, Kenya: ISAAA AfriCenter.

Kenya NPT regulations (2009). *Seeds and Plant Varieties (National Performance Trials) Regulations, 2009.* Available at http://extwprlegs1.fao.org/docs/pdf/ken37547.pdf. Accessed 14 March 2019.

Louwaars, N.P., Eaton, D.J.F., Tripp, R., Henson-Apollonio, V., Hu, R., Mendoza, M., Muhhuku, F., Pal, S. and Wekundah, J. 2005. *Impacts of strengthened intellectual property rights regimes on the plant breeding industry in developing countries.* Washington, DC: The World Bank.

Minde, I. 2004. *Harmonizing seed policies and regulations in Eastern Africa: experiences and lessons learned.* Entebbe, Uganda: Eastern and Central African Programme for Agricultural Policy Analysis.

Minde, I. and Waithaka, M. 2006. Rationalization and harmonization of seed policies and regulations in Eastern and Central Africa: effecting policy change through private public partnerships. *IN: International Association of Agricultural Economists Conference, Gold Coast, Australia, August 12–18, 2006.*

Munyi, P. and De Jonge, B. 2015. Seed systems support in Kenya: consideration for an integrated seed sector development approach. *Journal of Sustainable Development,* 8(2), pp. 161.

NSP. 2010. *Republic of Kenya: National Seed Policy,* Ministry of Agriculture: Nairobi.

Odame, H. and Muange, E. 2011. Can agro-dealers deliver the green revolution in Kenya? *IDS Bulletin,* 42(4), pp. 78–89.

Rangnekar, D. 2014. Geneva rhetoric, national reality: the political economy of introducing plant breeders' rights in Kenya. *New Political Economy,* 19(3), pp. 359–383.

SPVAA. 2012. *Seeds and Plant Varieties (Amendment) Act, 2012 (No. 53 of 2012).* Nairobi: Kenya Gazette Supplement, No. 217.

STAK. 2007. *STAK News.* 11/06 (2), pp. 1–5.

Tansey, G. 2011. Whose power to control? Some reflections on seed systems and food security in a changing world. *IDS Bulletin,* 42(4), pp. 111–120.

The Constitution of Kenya. 2010. *The Constitution of Kenya, 27 August 2010,* available at: http://extwprlegs1.fao.org/docs/pdf/ken127322.pdf. Accessed 14 March 2019.

USDA Foreign Agricultural Service. 2008. *Kenya planting seed report.* GAIN Report number: KE8010.

Waithaka, M. and Kyotalimye, M. 2013. *Harmonizing seed policy in Eastern and Central Africa* [Online]. Available from: http://knowledge.cta.int/Dossiers/ S-T-Issues/Seed-systems/Feature-articles/Harmonising-seed-policy-in-Eastern-and-Central-Africa-lessons-from-a-public-private-partnership-model. [Accessed 28th April 2016].

Waithaka, M., Nzuma, J., Kyotalimye, M. and Nyachae, O. 2011. *Impacts of an improved seed policy environment in Eastern and Central Africa* [Online]. Available from: https://www.asareca.org/~asareca/sites/default/files/publications/ Impactsofanimprovedseedpolicyenvironment.pdf. [Accessed 2nd June 2016].

World Bank. 2013. *Agribusiness indicators: Kenya.* Washington, DC: Agriculture and Environment Services, The World Bank.

CHAPTER 5

Ethiopia: A Transformationalist Seed Law

This chapter examines the exercise of seed sovereignty in Ethiopia. It focuses on the country's most recent Seed Proclamation 782/2013, which was signed into law by President Girma Wolde-Giorgis on 15 February 2013. This Proclamation was the first domestic seed law since Seed Proclamation 206/2000. It was also, significantly, the first seed law since the World Bank's[1] proposal for a far-reaching Agricultural Growth Programme (AGP) in the country in 2009[2] that ran in tandem with the Ethiopian Government's own growth programme,[3] and the establishment of the Ethiopian Agricultural Transformation Agency (ATA) in December 2010 by Federal Regulation 198/2010.[4]

---

[1] World Bank 2010 AGP Proposal: http://documents.worldbank.org/curated/en/63493 1468036896288/pdf/532900PAD0REPL1Official0Use0Only191.pdf [accessed online 16 August 2016].

[2] AGP which was officially launched in 2011 was designed "specifically targeting the Ethiopian seed system through technical support and investment" (Alemu 2011, p. 70) and was deeply connected with commanding a new direction for the multilateral donor agencies in tandem with the Ethiopian Government's own new GTP.

[3] GTP is the successor to previous government programmes Sustainable Development and Poverty Reduction Programme (2002–2007) and Plan for Accelerated and Sustainable Development to End Poverty (PASDEP) (2006–2010), which indicated a shift towards a market-economy and private-sector inclusion.

[4] The ATA emerged out of a "two-year extensive diagnostic study of Ethiopia's agriculture sector, led by the Ministry of Agriculture and facilitated by the BMGF" (ATA 2015). This

© The Author(s) 2019
C. O'Grady Walshe, *Globalisation and Seed Sovereignty in Sub-Saharan Africa*, International Political Economy Series, https://doi.org/10.1007/978-3-030-12870-8_5

156 C. O'GRADY WALSHE

The basic questions are the same as those in the Kenyan case study. Who wrote the Seed Proclamation? How was it drafted? What was the motivation behind the content of the Proclamation? Who was included in the process of determining its contents? Who was excluded? The purpose of asking these questions is to identify the main actors and events that led to the Proclamation. The overall aim is to understand whether a hyperglobalist, sceptical or transformationalist approach best captures the exercise of seed sovereignty in Ethiopia in the context of the preparation and passage of this Proclamation.

To answer these questions, I engaged in an examination of all of the relevant seed legislation and policy documents, but with greater reliance this time on personal contacts within Ethiopia, as much of the relevant legislative documentation was not available online. Once again the IDS Bulletin (2011) special edition on "The Politics of Seed in Africa's Green Revolution", proved useful and included Dawit Alemu's (2011) article "The Political Economy of Ethiopian Cereal Seed Systems: State Control, Market Liberalisation and Decentralisation". Alemu's paper provided a useful signpost to pursuit of further documentation and key informants, including the author himself. This contact proved invaluable. This chapter also benefits from the connection following intensive online research, to the work of Dr. Melaku Worede and Regassa Feyissa, both of whom became key Ethiopian contacts throughout this case study. Dr. Melaku provided clarifications on important aspects of seed policy and practice in the early stage of the research, through lengthy telephone interviews and later through a number of meetings. This clarified much of the context of the historical work on seed within the country and the important contextual setting regarding biodiversity, agroecologies, food insecurity and the complexity of the seed system.

The chapter is in three sections. The first section examines the main differences between the 2000 and 2013 Seed Proclamations. The second section identifies the regulatory roadmap leading to 782/2013, noting the key fora where critical changes occurred during an intensive drafting process. The third section focuses on the motivations of the main internal and external actors involved in the decision-making process.

new body was subsequently funded by BMGF (15%), alongside The World Bank (AGP/ Global Agriculture and Food Security Program (GAFSP)—26%), the Royal Netherlands Embassy (21%) and the Department of Foreign Affairs, Trade and Development (DFATD)— Canada (10%), United Nations Development Programme (UNDP) (8%), USAID (5%), GoE (5%) and others (ATA 2015, p. 105).

ETHIOPIA'S NEW SEED ACT: 782/2013

Whilst Seed Proclamation 782/2013 is a more succinct document than Seed Proclamation 206/2000, much of the text of Proclamation 206/2000 is dropped from the new one, while significant text is added in important areas of interest. Seed Proclamation 782/2013 differs in four key areas from its predecessor. They are in context, definitions, scope and authority. It is immediately evident that there are contradictory elements to the new law, which is significant in the context of my study.

1. Context: 'Improved' Seed for Market

The opening preamble of 782/2013 states "Whereas it has become necessary to facilitate the introduction of improved seed varieties to the market". This immediately sets the tone and intent of 782/2013, as improved seed is understood to mean private or company seed, as opposed to farmers' seeds or local seeds (GRAIN 2008). This reveals how the law seeks to bring Ethiopia's seed sector in line with commercial interests. All references in the preamble of 206/2000 to the diverse agroecological zones of Ethiopia, the protection and control of the originators of plant seeds as well as local adaptation of crop species and agricultural livelihoods have been removed.

2. Changes in Key Definitions

Seed Proclamation 782/2013 contains 21 definitions relevant to the entire new Proclamation, compared to the definitions in the 206/2000 Proclamation. However, many of the 206/2000 definitions have been dropped from the 782/2013 version and important new definitions added. In total, ten of Seed Proclamation 206/2000's definitions have been dropped and eight definitions have been significantly changed, while only one definition remains unchanged. There are now a whole plethora of new definitions added in Part One, Article 2, namely for breeder seed (Article 2.7), pre-basic seed (Article 2.8), basic seed (Article 2.9), modified organism (Article 2.11) and Quality Declared Seed (Article 2.13). These are significant additions, signalling a move by the Ethiopian Government towards opening up potential new seed practices/pathways, which would affect seed sovereignty in different ways. Most significant is the choice taken to adopt the UPOV 91 definition of what constitutes a 'variety'. This is a key feature of globalising seed law based on the strictest international seed regime, signalling a decision for closer alignment of Ethiopia's formal (albeit public) seed sector with global commercial markets of world technology and plant breeding.

UPOV Definition of 'Variety' In Seed Proclamation 206/2000, 'Variety' is defined as a "sub-division of any kind of plant species that can be clearly differentiated from other varieties of that kind by heritable characters; and that remain stable when reproduced sexually or asexually" (SP 206/2000, Part One, Section 2.3). In Seed Proclamation 782/2013, this is entirely changed to conform verbatim to the most recent definition of UPOV 91,[5] when it defines variety as follows:

'Variety' means plant grouping within a single botanical taxon of the lowest known rank that can be:

(a) defined by the expression of the characteristics of a given genotype or combination of genotypes;
(b) distinguished from any other plant grouping by the expression of at least one of the said characteristics; and
(c) considered as a unit with regard to its suitability for being propagated unchanged; (782/2013, Part One, Article 2.15)

This is a significant insertion, as the UPOV 91 definition is not legally required by Ethiopia as it is an exempt country under the UN-designated 'Least Developed Country' (LDC) status, whilst the Trade-Related Intellectual Property Rights (TRIPS) Agreement does not require countries to conform to UPOV standards until at least 1 July 2021 or beyond[6] (Munyi et al. 2016; De Jonge 2014, p. 101). Nevertheless, 782/2013 (Article 2.15) embraces the UPOV 91 definition, which demands that seed/plants pass the DUS test—that is display that they are distinct, uni-

---

[5] Explanatory Notes on The Definition of Variety Under the 1991 Act Article 1 (VI) of the UPOV Convention, adopted by the Council at its 44th ordinary session on 21 October 2010.

[6] Based on exemptions for LDCs dating back to 2003, but confirmed now in most recent Council for Trade-Related Aspects of Intellectual Property Rights Decision IP/C/64 of 11 June 2013.

form and stable, thereby fulfilling the requirements for the grant of breeders' rights, which is not otherwise explicitly mentioned in this new Seed Proclamation at all. Yet DUS and plant-breeders' rights (PBRs) are inextricably linked to homogeneity in plant systems and harmonisation and uniformity in the legal and commercial systems that accompany the new proprietorial rules over seed, which directly impacts on the subject of enquiry—Ethiopia's seed sovereignty. On top of this, the adoption of this new definition of 'variety', where it states "combination of genotypes" (Part One, Article 2.15 a), critically covers synthetic varieties and hybrids, which further embeds Ethiopia's seed future with biotechnological innovations and hybrid seed development, all associated with commercial/contractual agricultural methods, as opposed to farmers' seeds and farmer varieties.

Yet, 782/2013 shows greater flexibility than emerging seed legislation in neighbouring jurisdictions in its newly established recognition of two classes of commercial seed—certified seed and quality declared seed (QDS) (782/2013, Part One, Articles 2.10 and 2.13). This is in fact a marked change from Seed Proclamation 206/2000, which had a stricter certification standard 'for all crops'. 782/2013 gives a clear recognition to alternative seed pathways, with differing degrees of standardisation and certification, one, (certified seed) which must adhere to the new strictest definition and criteria, while the new (QDS) requires more relaxed quality standards.

Certified Seed Similarly, critical changes were made to the definition of what constitutes "Certified Seed". Heretofore in SP 206/2000 the definition was straightforward. Some seed was certified for standardisation purposes. It was not linked to proprietorial rights of PBRs and was defined as follows:

'Certified Seed' means a seed produced in Ethiopia or imported seed which has been certified, by the Agency or other body delegated by the Agency, for conforming to the standards established and which is intended for planting (SP 206/2000, Part One, Article 2.6).

However 782/2013 redefines what is meant by certified seed. Now "certified seed is contract seed" (Atilaw 2010, p. 23). It explicitly opts for conformity with the international seed regulatory framework, in this case relating to the definition of breeder as directed by UPOV 91, Article 1 (iv). It states that:

"'Certified seed' means a direct descent seed from basic seed or a seed found in first, second and third generation of basic seed" (782/2013, Part One, Article 2.10).

Certified seed is therefore linked back to the now newly added definition of breeder seed. Article 2.7 states: "'breeder seed' means seed of the first generation of seed multiplication, produced under the direct control of the breeder or his assigned representative" (782/2013, Part One, Article 2.7). This definition of certified seed accommodates new actors/ commercial interests where seed can be grown by "selected seed growers under the supervision of the seed enterprise, public or private" (ibid.). This is also critically linked to patent rights, and those with a commercial interest in developing new varieties. In Ethiopia's case, this is significant in the context of its considerable genetic resource wealth.

Quality Declared Seed Under 782/2013 Quality Declared Seed (QDS): is defined as "seed produced by organised and registered smallholder farmers or registered small holder farmers, in conformity with the required quality standards" (782/2013, Part One, Article 2.13). This is a significant new inclusion and indicates that Ethiopia is revising its law with a multi-tiered certification system in mind, as quality declared seed is a different seed category from strictly certified seed. QDS, while still subject to government-enforced regulation, does not require the same level of testing and quality standards as the purely formal seed sector requires. This is a significant inclusion. It highlights an insistence by Ethiopia at this time to ensure alternative seed pathways in such food-insecure locales, which presently rely almost entirely on diverse non-commoditised seed systems and broader heterogeneity in their seed varieties. It is also significantly different from Kenya's Seed and Plant Varieties Amendment Act (SPVAA) 2012, which makes no such provision despite having more than 80% of their smallholder farming population relying on similar non-commoditised systems of seed exchange (Munyi and De Jonge 2015).

Modified Organisms A new definition is also added with regard to "Modified Organisms" under 782/2013, signalling the potential realisation of another new pathway for seed practice in the Ethiopian context. This is significant largely because of the global seed transnational corporation (TNC) interests in the region and their objective of creating a legal framework to open Ethiopia and the region/continent to conventional biotech 'products' or seeds, a pathway, which Ethiopia has been at the forefront of opposing to date.

782/2013 defines modified organism as follows:

> 'modified organism' means any biological entity which has been artificially synthesised, or in which the genetic material or the expression of any of its traits has been changed by the introduction of any foreign gene or any other chemical whether taken from another organism, from a fossil organism or artificially synthesised. (782/2013, Part One, Article 2.11)

Elsewhere in 782/2013, provision is made for the importation of genetically modified organisms (782/2013, Part Four, Article 17.3), with special mention of consent being given if the "Ministry receives prior assurance of compliance with the applicable legislation from the Environmental Protection Authority" (782/2013, Part Four, Article 17.3). This appears to represent a loosening of arrangements around seed mobility, especially regarding controversial seed technology, particularly as Ethiopia is/was an ardent opponent of genetically modified organisms (GMOs) on the continent, notwithstanding its own global importance as a 'centre of origin' for key crop varieties/agrobiodiversity. The flexibility of the new seed law leaves it open to either restrict or enable such developments. However, it does maintain a level of control by insisting on Environmental Protection Agency (EPA) approval, ensuring developments in GE seed technology are kept firmly in government hands.

3. Scope of the Application

The adoption of globalised UPOV 91 definitions in key areas signals Ethiopia's intention to open up its hitherto closed seed arena to globalised seed pathways. This greatly favours and accommodates large commercial actors in the seed space. That said, an important insertion in 782/2013 seems to starkly contradict this assertion. This pertains to Part 1, Section 3.2 under the heading 'Scope of Application'. This gives more direct recognition to the smallholder farmers, by way of a critical exemption from compliance with other certification requirements found elsewhere in 782/2013. It states:

This Proclamation may not be applicable to:

(a) The use of farm-saved seed by any person;
(b) The exchange or sale of farm-saved seed among smallholder farmers or agro-pastoralists;
(c) Seed to be used for research purposes; and
(d) Forestry seed.

Points (a) and (b) here are more explicit than Proclamation 206/2000, which had stated that:

> this Proclamation shall not apply to a seed produced by a farmer, and sold directly to another farmer; However, this Proclamation shall apply to any producer or farmer, processor, distributor and retailer who advertised seed to engage in a sale of seeds. (206/2000, Part One, Article 3.2)

782/2013 exempts the majority smallholder farming population (97%) from compliance with the strict rules system, which is being applied for selected commercial crop varieties.

The inclusion of a much clearer exemption in the revised Seed Proclamation 782/2013 is unusual in the present tranche of 'enabling' legislative seed changes occurring on the continent. It is in stark contrast to the more stringent contemporaneous Kenyan law SPVAA 2012, which invoked the so-called farmers' privilege, a UPOV 2009 derivative term, with its inclusion of the phrase "within reasonable limits and subject to safeguarding the legitimate interests of the breeder". This term was designed to protect breeders by restricting farmers' rights to sell a commercial seed without recompense through royalty payment to the breeder. The new Ethiopian SP 782/2013 uses no such terminology and the exemption appears unequivocally in favour of the informal seed system and the farmers who rely on it for food security and income generation. This signals a stronger role for the Government of Ethiopia, or at least the ruling Ethiopian Peoples' Revolutionary Democratic Front (EPRDF)[7] party in government. In doing so, it establishes itself as a more powerful

---

[7] Following the election in 2010, EPRDF controls 99.6% of the House of People's Representatives (HoPRs), the highest legislative authority, which allowed them to form and lead the executive, the Council of Ministers and the Prime Minister, thereby allowing EPRDF to control both the executive and legislative wings of government (Lefort in Hassena et al. 2016, p. 93).

domestic/state actor in relation to its sovereignty over its seed, whilst simultaneously maintaining seed sovereignty in the hands of the majority smallholder farming population.

4. Authority

The 'Ministry', meaning the Ministry of Agriculture, is firmly established as the designated authority in the new seed law, at Part One, Article 2, point 16 (Federal Negarit Gazette No. 27, 15 February 2013). This is a change from 206/2000, which recognised the 'Board' of the National Seed Industry Agency[8] as the main competent authority at Part One, Article 2.1 (Proclamation No. 206/2000). Seed Proclamation 782/2013 brings every aspect of Ethiopian seed practice under Federal and Regional Governmental control. All powers in relation to seed variety release and registration (Part Two, 782/2013), integrated production planning and distribution of seed (Part Three 782/2013), quality control, certification, import and export, labelling, non-conforming seed, GMOs (which must also adhere to Ethiopian Environmental Protection Agency legislation), emergency seed (Part Four 782/2013), certifications of competence (Part Five 782/2013), as well as all matters relating to seed inspection, grievance procedures and offences and penalties (Part Six, 782/2013) are all now vested in the Ministry of Agriculture. Significantly, the section on National Variety Release and Registration in Part Two, Articles 4 and 5 of 782/2013 places sole responsibility and control in the hands of the 'Ministry', whereas 206/2000 recognised the 'Release Committee' as the competent authority in the approval, naming and registration of varieties (Part 1, Article 4, 206/2000). In this regard, 782/2013 requires variety registration for all crops, but is open to interpretation how it is to be done. This is significant as it is a critical area for the liberalisation and commercialisation of the private seed sector, and is linked to accreditation systems of a globalising seed framework.

Regarding the important area of import and export controls, the 206/2000 Proclamation had asserted that imported seeds must be of a

---

[8] The National Seed Industry Agency was dissolved into the Ministry of Agriculture and Rural Development (MoARD) by Proclamation 380/2004, Article 5, whereon MoARD became responsible for the promotion and expansion of agricultural development and to supervise all governmental and non-governmental organisations associated with seed production, distribution and regulation. MoARD at the federal level and BoARD at the regional level were given the responsibilities of implementing the seed policies and regulatory issues.

variety registered in Ethiopia (206/2000, Article 25) and that seed exports must be of varieties approved in Ethiopia and must meet Ethiopian quality standards (206/2000, Article 14). These stipulations are relaxed in 782/2013, Article 17, in the case of imports, giving the Ministry latitude in issuing future directives in this regard, while Article 27 copper fastens this discretion, adding an additional layer of ministerial power in Part Six, Article 27.2. This states that "The Ministry may issue directives necessary for the implementation of this Proclamation and regulations issued under sub-article (1) of this Article" (782/2013, Part Six, Article 27). Heretofore, this had been assigned to the National Seed Industry Agency (206/2000, Part Five, Article 35.2). This power to issue Regulation and directives gives the Ministry of Agriculture (MoA)/Government of Ethiopia considerably more powers directly relating to future regulatory functions and controls of seed practice.

Decentralisation of Power to Regional Authorities Seed Proclamation 782/2013 both defines the region and the regional authority in its key areas of definition (Part One, Articles 2.18 and 2.19). It then goes on to allocate a specific role for the implementation of national seed regulations to the regional authority, a new feature not included in 206/2000, which though it provided for delegation of authority did not specify any agency or body. This devolution or decentralisation of authority to the regional authority is in line with the Ethiopian constitutional[9] "commitment to a decentralised political-administrative system", according to Dawit Alemu of the Ethiopian Institute of Agricultural Research (EIAR) (Alemu 2011, p. 74). Its inclusion signals some new possibilities and challenges for regional autonomy and authority in the seed sector, despite the dominance of the centralised state. This has implications for seed sovereignty, because whilst the Ministry will be solely responsible for the aforementioned tasks, the regional authority will now have implementing authority, through contacts with companies, growers and dealers. 782/2013 devolves power to the regional authority for seed-testing laboratories (Article 11), seed quality control (Article 12) and the issuance of certificates of seed quality (Article 13) (Part Four, Articles 11, 12, 13, Seed Proclamation 782/2013). This devolution marks a significant new division of labour between the Ministry and the regions, which, already contentious, has the potential to impact on seed practices and therefore seed sovereignty.

[9] The "Constitution of the Federal Democratic Republic of Ethiopia, Proclamation No. 1/ 1995".

This section has highlighted the critical features of the new Seed Proclamation and provides clear examples of changes made, revealing some underlying ambiguities and many contradictory aspects to the emerging seed arena under a new legal framework 782/2013. It appears to give all the actors involved something that they are looking for. It is by no means a fully globalised law like the Kenyan SPVAA 2012 law, nor indeed recent laws in other jurisdictions within sub-Saharan Africa.[10] Whilst it embraces certain definitional aspects of IO UPOV 91, it by no means adheres to the strict bind of such a legal framework. 782/2013 opts for a differentiated seed system reflected through its insistence on recognition being given to three distinctly different levels of seed quality, access and farmers' rights. Certified seed for the formal seed system, quality declared seed for the commercial but less stringent seed system and a total exemption for its smallholder farming population allowing their informal seed system of exchange and sale to continue without deference to plant-breeders' royalties and the restrictions imposed in other jurisdictions on the continent at this time. Despite the perceived magnanimity of this acknowledgement, 782/2013 nevertheless is a commercially/market-driven and inspired seed law, giving vast powers of authority to the Government of Ethiopia (GoE), and by extension, the regional authority. This complicates things right down to the practical application of the law at the local district/worede and village/kebele level. Whilst it allows the farmer's exemption, it also allows giant transnational seed actors a clear mandate to pursue wide-ranging commercial seed practices that can significantly and immediately impact on the existing practices of seed sovereignty by Ethiopian farmers and communities.

The next section looks more closely at how 782/2013 came about, and the main drivers behind initiating new seed legislation.

THE STORY OF THE NEW SEED PROCLAMATION 782/2013: 2006–2013

There were five phases to the passage of the new Seed Proclamation 782/2013. The first phase began with the passage of two domestic Proclamations passed by the Ethiopian Parliament in February 2006, and

---

[10] Tanzania and Uganda are also in the process of fully enacting UPOV style laws, despite the LDC waiver until 2021.

coinciding with the arrival of Alliance for a Green Revolution in Africa (AGRA) and its Programme for Africa's Seed System (PASS) in the region. This was followed by the issuance of a 'Concept note' on Integrated Seed Sector Development (ISSD) in 2008, which established important state seed policy and principles. These were important signals of intent from the domestic state, which is entirely dominated by the ruling party, the Ethiopian People's Revolutionary Democratic Front (EPRDF). They directly preceded the initiation by the Ministry of Agriculture of a review of Seed Proclamation 206/2000 in February 2008, when the formal drafting of a new Seed Proclamation began. This phase produced the first two drafts of the new seed law. The second phase saw the intervention of two new external actors working 'collaboratively' with the Ministry of Agriculture, namely the International Development Law Organisation (IDLO) and the Dutch Embassy. This phase produced another two drafts. The third phase marked the publication of Draft Four and was a significant milestone, as it was at this point that Bill and Melinda Gates Foundation (BMGF)-inspired and part-funded new ATA took over the entire drafting process. Phase four saw the reworking of the MoA/IDLO/Dutch Draft by ATA's technical team. Following consultation 'on every Article and paragraph', ATA produced draft 5 and finally draft 6. The fifth and final phase of the process is when the draft is returned to the domestic legislative arena and passed through the various necessary parliamentary and legislative steps until its final passage into law in February 2013.

Figure 5.1 lays out a roadmap for the passage of the law. The major state/domestic decisions, proclamations and initiatives are represented on the left hand side of the map. The relevant seed interventions on behalf of international organisations (IOs), intergovernmental organisations (IGOs), international institutions and transnational actors are represented on the right hand side. While some of the external actors/actions collaborate with the domestic state through other organs or institutions and thus straddle both sides, I nevertheless separate them for the purpose of clarifying their status as either internal or external actors in this case.

The Road Map to 782/2013: The Preparation for the New Seed Proclamation

Phase 1: The Legislative Backdrop to 782/2013
Two domestic Proclamations relevant to the later passage of 782/2013 were passed by the Ethiopian Parliament on 27 February 2006, namely (a)

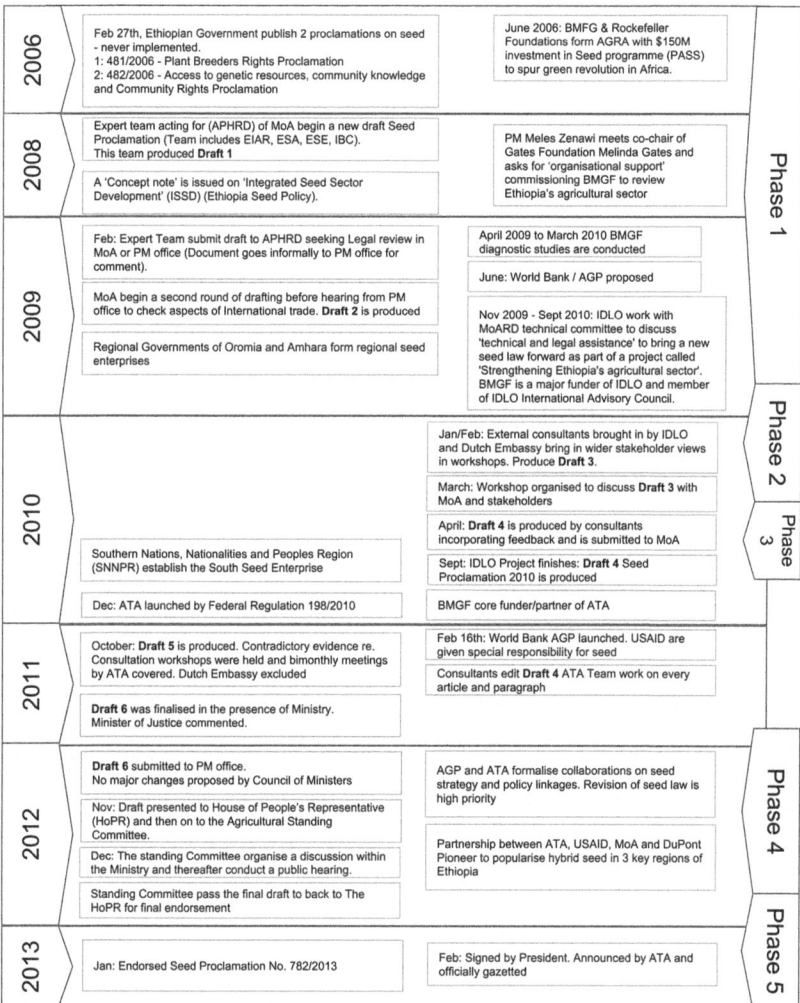

Fig. 5.1   The road map to SP 782/2013: chronology of key events 2006–2013

the plant-breeders' rights (PBRs) Proclamation (481/2006)[11] and (b) access to Genetic Resources and Community Knowledge, and Community Rights Proclamation (482/2006).[12] That Ethiopia instituted such Proclamations just prior to the deepest penetration of its seed systems by global actors is significant.[13] Though 481/2006[14] remains to be fully implemented by parliament (Feyissa 2006, p. 7; IDLO 2013), both are influenced by globalising laws. PBRs legislation (481/2006) is associated with the strictest international seed standards, namely UPOV. 482/2006 is associated with farmers' rights and access and benefit sharing (ABS) enshrined in the Convention on Biological Diversity (CBD) and the International Treaty on Plant Genetic Resources for Food and Agriculture (ITPGRFA), which Ethiopia is party to.[15] The focus of CBD is access and benefit sharing based on individually tailored ABS agreements through national sovereignty of genetic resources, while ITPGRFA focuses on international pooling and sharing of genetic resources through a multilateral system of access and benefit sharing (Bioversity International 2015). However, Ethiopian law 481/2006 did not follow the straightforward UPOV template for a PBR legislative framework. It pointedly included the role of 'local farmers' and their contribution to conservation and use of genetic resources. Feyissa (2006) states that this "constitutes the basis

[11] Federal Negarit Gazeta of the Federal Democratic Republic of Ethiopia. Vol. 12 No. 12, 27 February 2006, pp. 3339–3352. 481/2006 was not implemented and no Plant Variety Protection (PVP) legislation has been adopted by Ethiopia to date. A new PBR Proclamation, which has been reworked by various other international actors, namely IDLO who state that given the important implications of this legislation, it "has designed and will implement this component in partnership with Wageningen University of the Netherlands, which will be responsible for technical aspects of breeders' rights", (IDLO 2013) [accessed online 2/11/2013].

[12] Federal Negarit Gazeta of the Federal Democratic Republic of Ethiopia, Vol. 13, 27 February 2006, pp. 3353–3373.

[13] ASARECA and EASCOM pilot project had been extended to Ethiopia and other countries in 2001 and through the Seed Regional Working Group (S-RWG) the push for harmonised seed legislation under COMESA was well in train (Waithaka et al. 2011, p. 8). In June 2006 AGRA launched its 'Green Revolution for Africa' programme in East Africa with plans for a huge onslaught on changing seed systems through its PASS programme.

[14] According to van den Broek, an entirely new PBR Proclamation instead has been submitted to parliament and is presently in the final stages of ratification by the Council of Ministers. The draft law strikes a balance between strong breeders' rights for horticultural crops (vegetables and ornamentals) while having greater farmers' rights for crops that are important for food security (cereals and legumes) (Broek 2015, p. 19).

[15] Article 15 of 482/2006 provides for a special access permit in fulfilment of ITPGRFA.

for breeding new varieties for agricultural production" (Article 27) (quoted in Feyissa 2006, p. 6), which may be the reason for the delay in its passage, as this was precisely the nature of the tension surrounding who would have sovereignty over Ethiopia's seeds. The final version emerging from the Council of Ministers' deliberations will be revealing in that regard. Similarly, Proclamation 482/2006 was designed to "ensure that the state and communities obtain fair and equitable shares from the benefits arising from the use of genetic resources", establishing that "the state decides on access to genetic resources, while communities decide on access to community knowledge" (Article 5, 482/2006). This Proclamation designated the Institute of Biodiversity Conservation (IBC), now called Ethiopian Biodiversity Institute (EBI) (itself formally established by Proclamation 120/1998), as the institution which would have the power "to decide on and facilitate access to genetic resources and community knowledge" (Feyissa 2006, p. 7). EBI is one of the main 'Accountable Institutions' of state and would be a key actor on the first expert team appointed by the Ministry of Agriculture to redraft the new seed law. It is also significant, as we see in the following sections, that some of these principles and bodies such as EBI were included in the draft Seed Proclamation of 2010, yet not all were maintained in the final Seed Proclamation 782/2013 three years later when other external actors entered the drafting process.[16] In this way, these Proclamations provide us with an insight into the Government of Ethiopia's position and how it changed over the course of the years ahead.

Draft 1 (2008) and Draft 2 (2009) of New Seed Proclamation
In late 2008, the Ministry of Agriculture initiated a process of drafting a new Seed Proclamation. This coincided with the first 'fortuitous' introductory meeting between Prime Minister (PM) Meles Zenawi and Melinda Gates of BMGF,[17] at which a commitment was allegedly sought and given for BMGF to assist in a review of Ethiopia's agricultural sector. Whether or not the Gates meeting in 2008 instigated the move is not clear. However, BMGF had already established projects/interests in Ethiopia since 2000 and would eventually become centrally involved at many levels, but mainly through the establishment of a new agency, ATA, in 2010.

---

[16] In fact, any mention of EBI was entirely dropped from the final Seed Proclamation 782/2013.
[17] http://www.ata.gov.et/about-ata/origin-history-2/ [accessed online 18/8/2016].

ATA thereafter became solely responsible for producing the final two drafts of the Seed Proclamation. In 2008, preceding this period, the drafting process was led by the Animal and Plant Health Regulatory Directorate (APHRD), 'a bureaucratic department' in charge of seed regulation within the Ministry of Agriculture (Hassena et al. 2016, p. 87). MoA began by bringing together a team of experts, notably the Ethiopian Institute of Agricultural Research (EIAR), Institute of Biodiversity Conservation (IBC/EBI), Ethiopian Standards Agency (ESA) and Ethiopian Seed Enterprise (ESE), all key domestic actors as public institutions with pivotal roles as stakeholders in the seed space at a federal level (Interview with Ayenew, 12 November 2015). This is the expert team that reviewed the documentation and produced the first draft of a new Seed Proclamation. APHRD did not include other stakeholders or external actors (Hassena et al. 2016, p. 87). In their study, Hassena et al. (2016) reveal the turf war between public and private interests at the outset. They also firmly establish that the Ministry in general and the Prime Minister's office and experts in particular were centrally involved from the beginning in determining certain outcomes. This was particularly the case with regard to "the major emphasis placed on food crops, whilst the issue of export crops (horticulture in particular) was not considered" (ibid.). This 'emphasis' on smallholder farmers and their seed/agricultural practices which emerges here in the first drafts, emanating as it did largely from the domestic state and public institutions, became an important and defining feature of the subsequent Seed Proclamation 782/2013. This is explained in greater detail below through interviews with the various actors involved, notably MoA, EBI and EIAR. This emphasis finds its main articulation in the exemption afforded to smallholder farmers in favour of their practice of seed saving, exchange and sale, which remained throughout the entire drafting process, despite other additions in favour of globalising seed measures. It is significant as such an exemption was not afforded in the text of Kenya's contemporaneous seed law, nor in a number of other African jurisdictions undergoing similar legislative changes.

The expert team submitted a first draft of the new Seed Proclamation in February 2009, seeking legal review from the MoA or Prime Minister's office. According to Hassena et al., it was at this stage that a "document was sent informally to experts within the Prime Minister's office also to obtain their comments" (Hassena et al. 2016, p. 87). At the same time, the Ministry continued with the process of constructing a second draft. Meanwhile, some concerns had arisen regarding the application of the new law and international trading capability "triggered ... [by] the confiscation of Ethiopian flowers at Schiphol airport, in the Netherlands, at the end of 2008" (Hassena et al. 2016, p. 88), sparking major discussions between MoA and the Dutch Embassy in Addis over Intellectual Property Rights (IPRs) and seed policy (ibid., p. 88). However, the experts at the Prime Minister's office sent back a text dropping mention of any new office or the exemptions favouring exporters, just as the Ministry of Agriculture began working with an important Intergovernmental organisation, namely the International Development Law Organisation (IDLO), on bringing the next two drafts of the new Seed Proclamation forward.

Phase 2: Draft 3 and Draft 42,010
Following the alleged 'fortuitous' first meeting between BMGF and PM Meles Zenawi in 2008 and while several diagnostic studies funded by BMGF were being undertaken, another key external actor, IDLO[18] with close links to BMGF, became centrally involved in the drafting process of 782/2013. In 2009 IDLO's legal counsel approached Ethiopia's Ministry of Agriculture Legal Affairs Directorate, namely Daniel Mekonnen, proposing a new seed law (Interview with Mekonnen at Ministry of Agriculture and Rural Development, 11 November 2015; IDLO documentation [Accessed online 2/11/2013]). Hassena et al. (2016) assert that the Ministry approached IDLO and the Dutch Embassy at this time seeking support 'in drafting the new seed law', which they say was positively received and consultants were hired to draft it (Hassena et al. 2016, p. 88). This contradicts Daniel Mekonnen's account, where he states that "after drafting the Proclamation it needed to be based on international law", going on to say that:

IDLO came to our office and they wanted to work with us. They helped us draft the Proclamation in English, and their legal person was organising that. Then after that we summoned the Ministry, then the Minister of Justice commented and then the Prime Minister's office reorganised the

[18] IDLO describes itself as "the only intergovernmental organisation exclusively devoted to promoting the rule of law" (https://www.idlo.int/about-idlo/mission-and-history. Accessed 26 December 2018). Based in Rome since its formation in 1988, IDLO "enable governments, empower people and strengthen institutions to realize justice, peace and sustainable development" (ibid.). It counts among its funders the Bill and Melinda Gates Foundation as well as listing William Gates Senior, co-chair of BMGF as a member of its five strong International Advisory Council.

draft, before going to the Council of Ministers and on to the upper chamber, the House of People's Representatives. (Interview with Mekonnen, 11 November 2015)

Whoever made the first step, we do know that the meeting between IDLO and Ministry of Agriculture and Rural Development (MoARD) initiated a ten-month (November 2009–September 2010) project titled "Strengthening Ethiopia's Agricultural Sector", which produced two more drafts (3 and 4) of the Seed Proclamation, and which was carried out jointly between IDLO and MoARD (Interview with Mekonnen, 11 November 2015 and IDLO online 2013). Their expressed task was to "develop a new regulatory framework" based on the premise that "the legal and institutional framework for agriculture had become outdated and largely ineffective, in particular with respect to international standards and practices" (IDLO 2013). According to IDLO documentation, they thereafter identified four key areas along with MoARD, which they would work on as a priority: (1) seed legislation and regulations; (2) plant-breeders' rights legislation; (3) animal health legislation; and (4) plant protection legislation. The Project's first objective was to update Ethiopia's Seed Proclamation 206/2000. The consultancy team brought in a wider body of stakeholder opinion and submitted draft three to the MoA, suggesting a stakeholder workshop be organised, which duly occurred in March 2010, where "the issues of variety registration and seed sector governance became even more contentious" (IDLO cited in Hassena et al. 2016, p. 88). At the end of this period, the Ministry remained resolutely determined to keep seed governance within the Ministry, as the issue of an office of registration for varieties did not come through, whilst the exemption for registering certain export-only varieties was conceded and remains in 782/2013 (Article 17). The Ministry's resolve appears to have been emboldened by the returned document from experts in the PM's office, which had omitted both issues, yet the 'general policy privilege of the export sector' meant that this particular issue could not be opposed as strongly (ibid., p. 89). Following this workshop, a new draft Seed Proclamation was published in April 2010, draft four in our tracking process. This thereafter became the working document for further revision by key actors to follow, namely the ATA, which was established by Federal Regulation in December 2010, two months after the IDLO/MoARD draft Seed Proclamation came to fruition. IDLO immediately proceeded to work on a revision of the Ethiopian Plant Breeders' Rights Proclamation

No. 481/2006[19] "to strengthen and assist with its implementation",[20] in conjunction with Wageningen University of the Netherlands (IDLO online 2/11/2013)[21] while ATA now took centre stage and would bring Draft Seed Proclamation 2010 (draft 4) forward to completion. This is a crucial stage of the process, as it allows us to examine what changes have happened up to this point with the influence of the IDLO and Dutch Embassy involvement alongside the MoA, the PM's office and other domestic actors for the Ethiopian side.

Phase 3: Draft Seed Proclamation 2010
The draft Seed Proclamation brought forward by IDLO together with the Ministry of Agriculture's legal directorate in 2010 contained many changes to the 206/2000 Seed Proclamation and some additions which remained in the 2013 final Seed Proclamation 782/2013. This is a significant text[22] at a key moment in the drafting process. This is, firstly, because it gives us an insight into what influence IDLO and the Dutch Embassy may have come to bear on the process since the early drafts of 2008 and, secondly, because it allows us to examine the difference between this draft and the final draft six, when ATA have sole control over the process from this point on.

I identify four key differences between 206/2000 and draft SP 2010 below. They relate to (1) contextual changes; (2) the introduction of significant definitional changes; (3) the identification of a 'competent authority' regarding seed; and (4) the scope of the application of the new Seed Proclamation, with an important exemption for smallholder farmers.

It is at this point we note that all reference to agroecological zones of Ethiopia is dropped from the draft SP, which marks a significant shift. Yet, within a plethora of definitional changes we also see the inclusion of three standards for seed quality, namely Certified Seed, Quality Declared Seed and Emergency Seed (Part 1, Article 2:4), suggesting a facilitation for a

---

[19](Federal Negarit Gazeta of the Federal Democratic Republic of Ethiopia. Vol. 12 No. 12, 27 February 2006, pp. 3339–3352).

[20] However, 481/2006 was not implemented and no Plant Variety Protection (PVP) legislation has been adopted by Ethiopia to date.

[21] IDLO states that given the important implications of this legislation, it "has designed and will implement this component in partnership with Wageningen University of the Netherlands, which will be responsible for technical aspects of breeders' rights" (IDLO 2013).

[22] This text of Draft Seed Proclamation 2010 can be found as Annex 6 in Atilaw (2010).

differentiated approach within alternative agroecosystems. Draft SP 2010 specifically names the "Minister" or "Ministry", as the competent authority responsible for agriculture (Part 1, Article 2.7), with a note added by the drafters stating that "[t]his formulation is a useful alternative to putting in the current name of the Ministry, which if changed would necessitate an amendment of the Proclamation" (ibid.). Similarly "region" and "regional authority" are new definitional additions from 206/2000 Proclamation, signifying that perhaps an increased role is foreseen for them and once again the explanation that the wording is such that it allows variation on the name of the regional agricultural authority in the future without needing to amend the legislation (Draft SP 2010, Part 1, Article 2.14, 15). This is very significant in the context of the push to establish a national office by many external actors involved as they sought alignment with other seed regulatory bodies in the region. This is a major objective of the Association for Strengthening Agricultural Research in Eastern and Central Africa (ASARECA), Eastern Africa Seed Committee (EASCOM) and Common Market for Eastern and Southern Africa (COMESA) agendas, who seek to take seed regulation out of public/ministerial control, as they had just succeeded in doing with Kenya Plant Health Inspectorate Service (Kephis) in Kenya, opening it up to privatisation.

Other globalising features are introduced at this stage also with the commitment to UPOV definition of 'variety', stating that: "The UPOV definition has been used to bring the legislation in line with Ethiopia's Plant Breeders' Rights Proclamation and international standards" (ibid.), a signal of intent to develop new 'improved seed' for market access in line with the strictest IO framework.

However, paradoxically, and most significantly in the context of this study, it is draft SP 2010 Part One, Article 3, which established the critical exemption for smallholder farmers stating that:

This Proclamation shall not apply to:

(a) the use of farm-saved seed by any person, or the use, exchange or sale of farm-saved seed by smallholder farmers;
(b) any sale of seed involving persons or seeds exempted from the provisions of this Proclamation under sub-article 3; or
(c) seed used for research purposes, which nevertheless may be subject to regulation under other legislation. (Article 3.2)

Article 3.3 states that "the Minister may, from time to time and by published directive, exempt any class of persons or category of seed from the provisions of this Proclamation", which is entirely dropped in the final document 782/2013. However, the main exemption remained and was enacted.

In Part Three, Article 10.1 states:

> Varieties for export only shall not be subject to variety release or registration requirements unless otherwise required by the importing country, but shall nonetheless be subject to all controls under prevailing phytosanitary, bio-safety and other relevant legislation. (Article 10.1)

This corresponds to the assertion of Hassena et al. (2016) that this exemption was allowed as key horticultural exporters/stakeholders pushed for it.

The 2010 draft also notes the establishment of a National Variety Release Committee (Article 6) and a National Seed Advisory Board (Article 7), specifically naming the important domestic actor, IBC/EBI as a core member in both cases. Any mention of the IBC/EBI is entirely dropped from the 782/2013 final version. While Article 4.2 of 782/2013 still maintains the idea of a National Variety Release Committee, IBC/EBI does not feature and the other suggested Board is dropped entirely. This is extremely significant because EBI is the flagship organisation and steward of Ethiopia's coveted gene bank where its germplasm exchange with farmers is regarded as the essence of best practice for participatory plant breeding, climate resilient agroecological applications and seed sovereignty.

At this point, this draft was sent back to MoA and, as pointed out by Hassena et al. (2016), was expected to be sent on to the Prime Minister and passed by parliament, but "the draft was not submitted for more than a year" (Hassena et al. 2016, p. 89). They cite the May 2010 elections and subsequent changes in executives within the Ministry as possible reasons. However, this seems unlikely given what ensued a few months later as other mainly US actors were selected to take over the process. A more likely reason is that at this time the government was about to launch the ATA alongside the BMGF. This was the culmination of years of work behind the scenes. The new body, with the Ethiopian Prime Minister as its kingpin and Council of Ministers (CoM) in key positions, backed by a giant US foundation alongside predominantly US actors was about to take over the entire drafting process of Ethiopia's Seed Proclamation. By December 2010, two months following the completion of IDLO/MoA work on the document, ATA took over the seed drafting process (now draft 4) as its top priority and brought it to completion within two years.

Phase 4: The Establishment of ATA and Its Role in Changing Ethiopia's Seed Legislation

Following the alleged 'fortuitous meeting' which took place between Melinda Gates, co-chair of BMGF, and Ethiopian Prime Minister, Meles Zenawi, in 2008, where according to ATA, he proposed that BMGF provide 'organisational support' to Ethiopia (Interview with Bento at ATA 16 November 2015), the first BMGF commissioned study to write a review of agriculture in the country was set in train.

Pursuant to the BMGF-funded diagnostic studies, the Agricultural Transformation Agency (ATA) was established in December 2010 when the Council of Ministers passed Federal Regulation 198/2010.[23] The composition of the governing council of ATA is critical here. ATA would be governed by an 'Agricultural Transformation Council', with an executive committee chaired by the Prime Minister himself (firstly by Meles Zenawi and following his death in August 2012, by the then Prime Minister Hailemariam Desalegn). It consisted mainly of members of the ruling EPRDF cabinet/executive. The Minister for Agriculture (Tefera Derbew) was appointed as Vice-chair. The Transformation Council also included five other Cabinet Ministers, as well as the heads of the four main regional Bureaus of Agriculture, the Director General of the Ethiopian Institute of Agricultural Research (EIAR). Khalid Bomba, a member of the Ethiopian diaspora and former executive of the Gates Foundation, was appointed as the CEO. The ATA acts as the Secretariat to the new Council with a key objective "to identify systemic constraints of agricultural development" (Part Three, Article 9, Regulation No. 198/2010).

ATA immediately set about what it termed 'the Agricultural Transformation Agenda', a mechanism designed to provide a coordinated approach to remove the structural bottlenecks that constrain the achievement of specific agricultural targets (ATA 2015, p. 14). The first Transformation Agenda was composed of '84 deliverables' (ibid.). ATA

---

[23] "Agricultural Transformation Council and Agency Establishment Council of Ministers Regulation No. 198/2010."

identified the seed regulatory framework as one of "the systemic bottlenecks constraining development of the sector" and limiting the capability of the private sector (ibid., p. 45). The transformation of seed regulations and policies would be a first priority, starting with a revision of the Seed Proclamation.

Hassena et al. state that in mid-2011 the ATA began the process of redrafting the Seed Proclamation, relying solely upon ATA staff, ministry experts, Ministry decision-makers and researchers (Hassena et al. 2016, p. 89). This was confirmed by Yit Barek, the ATA official responsible for Seed law Development. He recalled that "key officials at ATA facilitated the drafting of 782/2013, by creating a platform to discuss the policies, and then organising a technical team to work on each Article and paragraph" (Interview with Yit Barek 11 November 2015). This was done through "a series of workshops, bi-monthly meetings and consultations with the Ministry of Agriculture as key partners" (ibid.). However, the Netherlands embassy and key members of the horticultural sector did not know about these meetings, according to Hassena et al. (2016), despite their involvement in the earlier drafts, signalling a significant shift in actors and influence from this point on.

Phase 5: 782/2013 Passage Through Ethiopian Parliamentary Process
Draft 6 was finalised by the Ministry of Agriculture and was commented on by the Minister for Justice at the end of 2011 (Interview with Mekonnen, 11 November 2015). "This draft was then submitted to the Prime Minister's office during the second half of 2012" according to Hassena et al. (2016, p. 90). It was then endorsed by the Council of Ministers and from there it went to the House of People's Representatives, the upper chamber[24] and highest legislative authority in November 2012 (Interview with Mekonnen, 11 November 2015). From here, the House of Peoples' Representatives (HoPRs) sent the draft to a standing committee of parliament for the relevant sector (in this case agriculture) where it was evaluated. Thereafter, the standing committee organised a separate discussion with the Ministry in December 2012, the purpose being "to build consensus between the Ministry and the standing committee", and to ensure common policy agreement between the executive and legislature (Hassena et al. 2016, p. 90). Hassena et al. (2016) note that both the

---

[24] This chamber is made up of 550 members elected each five years in a general election, of which 99.6% were returned for the ruling EPRDF party in the disputed 2010 elections.

HoPRs and the executive (Prime Minister and Council of Ministers) "have the power to draft policies but that the drafting of policies by the executive is mainly conducted by sectoral agencies" (ibid., p. 84). This is important as Hassena et al. (2016) assert that the ruling EPRDF party have assigned more power to the executive through stacking the executive with higher-ranking, senior members of the party, as well as the brightest and best members of the HoPRs, resulting in a politically and intellectually weakened legislature with an increased culture of dependency on the executive, but which is now also rendered virtually meaningless in the policy and law-making process (ibid.). From that disadvantaged point, the standing committee held 'a public hearing' to check if any issue had been left out. Hassena et al. (2016) note that while this hearing was open to the public, in fact "participation was limited to representatives of different government offices and the private sector" (ibid.). This included the Ethiopian Horticulture Producers Exporters Association (EHPEA). Some objections were raised at this stage by private-sector interests seeking an exemption for export-only varieties and this was conceded to and resolved, according to Hassena et al. (ibid.). An updated document was then passed back to the Ministry which then endorsed it without contestation. Thereafter, the standing committee passed "the draft back to the HoPRs for final endorsement" (Hassena et al. 2016, p. 90). Seed Proclamation 782/2013, was thereafter signed into law by President Girma Wolde-Giorgis on 15 February 2013. I now look more closely at the role of the key actors involved in the decision-making process and how they determined particular seed outcomes at key moments in the drafting process of 782/2013 in Ethiopia at this time.

THE KEY ACTORS BEHIND THE NEW ETHIOPIAN SEED LAW 782/2013

This section examines the role of the most influential domestic/state actors in bringing Seed Proclamation 782/2013 forward. Dominated by the EPRDF-run Government of Ethiopia, with its powerful executive, headed by the Prime Minister, it focuses on the role of the PM, who alongside experts in his office was central to the process of delivering a new seed law. I also examine the key role played by other important domestic actors, most notably the Ministry of Agriculture, and the crucial role and motivations of the main external actors, including other state actors, key IOs and transnational actors.

The process of revising the seed law was conducted by the APHRD, as the department responsible for seed regulation within the Ministry of Agriculture. It was they who, in February 2008, convened the first team of experts from the domestic state's own 'accountable institutions' with an interest in seed. This included The Ethiopian Institute of Agricultural Research (EIAR), Ethiopian Biodiversity Institute (EBI), Ethiopian Seed Enterprise (ESE) and Ethiopian Standards Agency (ESA). It is also significant that they did not host "a wider stakeholders consultation" at this time (Hassena et al. 2016, p. 87). Instead, they were clear to update the office of the PM both formally and informally of the outcome at critical stages of the drafting process, ensuring that it was very much a government-led and controlled exercise with the main Ethiopian state actors. It also clearly demonstrates that the MoA was following instructions from the Prime Minister's office throughout, as evidenced by the decision of APHRD to follow the revised version (which became draft 2) in 2009, following comments from the PM's office (IDLO 2010, quoted in Hassena et al. 2016, p. 88). Hassena et al. (2016) assert that the experts who were central to the early stages of drafting placed overdue "emphasis on food crops, while the issue of export crops (horticulture in particular) was not considered" (ibid., p. 87) at this point. However, some of these experts themselves express confusion regarding the motivations behind the drafting process and did not necessarily feel that their concerns were realised through 782/2013 (Interview with Feyissa, 26 November 2013). Regassa Feyissa, Director of EOSA,[25] former Director of EBI, a key domestic actor[26] in the seed space, recalls that there was already concern regarding moves into the Ethiopian seed space as far back as the early 1990s, when the World Bank gave Pioneer (now Du Pont/Pioneer)[27] a loan of $30 million to develop a private seed industry in Ethiopia. Moves by the Ethiopian Ministry and wider state actors at that time led to the establishment of an expert task force chaired by the Deputy Prime Minister and which included Feyissa, Dr Melaku Worede and Dr Tewolde Berhan

---

[25] EOSA—Ethiopian Organic Seed Action, an offshoot of SOS Seeds of Survival founded by Melaku Worede.

[26] Feyissa, a seed physiologist and biochemist, is one of Ethiopia's key experts at CBD and was successor to Worede at IBC/EBI.

[27] Pioneer Hi-Bred was the world's largest seed company when Du Pont took it over in 1999 (Kloppenburg 2014). Du Pont now holds 25–30% of the hybrid maize market share in Ethiopia (ACB 2015, p. 35).

Gebre Egziabher,[28] who was the head of Ethiopia's Environmental Protection Agency (EPA). The expert advice from these globally acclaimed scientists led to the formation of a seed policy, which enshrined farmers' rights and space for the farmer's role in varietal development in the Seed Proclamation of 206/2000. These earlier moves by the Ethiopian Government signalled a strong state attachment to maintaining seed sovereignty in both state and farmers' hands. Critical elements from this period continued to feature in the two later Seed Proclamations of 2006. Though these elements did not entirely survive the 782/2013 drafting/revision process, they provide some explanation for the distinct, albeit ambiguous state articulation of a differentiated seed paradigm that did emerge. Critically, this included the farmer's exemption (782/2013 Article 3.2 a), and the incorporation of a differentiated three-tiered seed categorisation (Part 1, Art 2.13), despite the pressure from globalising IP-dominated seed forces occurring across the region at this time (Louwaars et al. 2013). The EIAR representative in the drafting process, Dawit Alemu, describes this as Ethiopia's "peculiar style of Green Revolution that differs in important respects from other countries in the region" (Alemu 2011, p. 75).

Alemu, in both his 2011 paper on *The Political Economy of Ethiopian Cereal Seed Systems* for IDS Bulletin and in interview with him in 2013, highlights that despite the lack of capacity of the Ethiopian Seed Enterprise (ESE), the government retain tight control of market actors and all changes, as "tensions exist between the technocracy and the political system" (Alemu 2011, p. 7). In an interview with him he explains further, citing three defining features which determine this 'peculiar' application in the Ethiopian case. (1) State 'ambivalence to the private sector and economic liberalisation'. (2) A high degree of 'sensitivity to biosafety',

---

[28] Dr. Tewolde Gebre Berhan Egziabher was the Director General of the Ethiopian Environmental Protection Authority (EPA) until his retirement in 2012. An internationally acclaimed plant ecologist, he has been GoE main negotiator at international fora including on CBD, Food and Agriculture Organisation (FAO), Intergovernmental Panel on Climate Change (IPCC) and it was his leadership of the 'Like-Minded Group' in negotiations of the Cartagena Protocol on Biosafety which succeeded against strong US and EU opposition to protect biodiversity, human health and traditional and community rights. He is acknowledged as the key government expert on PGRs and seed legislative changes are generally examined by him. He remains a key advisor to the new Minister of Environment, Forest and Climate Change. He has won the Right Livelihood Award in 2000 and the UN top Environmental Prize in 2006 'Champions of the Earth'.

through a strong 'fighting' Environmental Protection Agency. (3) The very strong informal seed sector (Interview with Alemu 2 December 2013). He points to the continuation of "a top-down centrally-designed, state-directed approach to seed production and distribution. This system has been unchanged since the Imperial Regime, with state interest also dominating the formal seed system", according to Alemu (ibid.).

The nature of the drafting process from 2008 onwards and the differentiated outcomes in Ethiopia's new Seed Proclamation give an indication of where these tensions lay and an indication of what was won and lost and by whom. In this regard, it is worth noting the government's own, somewhat contradictory agenda. Despite their propensity for biological sensitivity on the one hand and ambivalence to the private sector on the other, the advancement of SP 782/2013 occurred within the backdrop of the Government of Ethiopia's ambitious Agricultural Development Led Industrialisation (ADLI) Strategy.[29] This intensified under the present strategy, namely Growth and Transformation Plan (GTP),[30] alongside Ethiopian attempts to join WTO.[31] These were all initiatives of the late Prime Minister Meles Zenawi and were central to his thesis on the "developmental state",[32] which sought a "paradigm shift in development thinking" (Interview with Ghidey Debessu, advisor to Minister of State for Agriculture, 25 November 2013). This big-vision, state-led programme marked an aggressive drive towards agricultural industrialisation and increased commercialisation. The new Seed Proclamation which included

[29] ADLI was initiated by the GoE in 1993 followed by 'A Plan for Accelerated and Sustained Development to End Poverty' (PASDEP) (2005–2010) and then the GTPs 2011–2015 (GTP1) and 2015–2020 (GTP2).

[30] GTP is the successor to previous government programmes Sustainable Development and Poverty Reduction Programme (2002–2007) and Plan for Accelerated and Sustainable Development to End Poverty (PASDEP) (2006–2010), which indicated a shift towards a market-economy and private-sector inclusion. The stated aim of GTP is to "foster broad based development in a sustainable manner to achieve the MDGs", whilst seeking to double the agricultural production of the country by 2015 and is considered centrally important to achieving middle-income country status by 2025.

[31] Ethiopia formally requested membership of WTO on 13 January 2003 as it attempts to transform itself into a market-oriented economy. Its request has not been successful to date (Asmelash 2014). http://papers.ssrn.com/sol3/papers.cfm?abstract_id=2523859 [accessed online 10/10/2016].

[32] "Developmental State is an institutional, political cum ideological arrangement that evolved from Japan's, post war economic recovery and was later adopted by some East Asian countries." The EPRDF started to articulate this concept in the early 2000s.

core changes to seed practice became a defining moment in this new 'transformation agenda'.

Alemu of EIAR further explained the government's thinking and actions at the time.

> You have a national framework (i.e. GTP)—to double agricultural production by 2015. Everyone is running against that target and 'improved technology' is the criteria of evaluating your comparative advantage in achieving that. (Interview with Alemu, 2 December 2013)

The introduction of 'improved seeds'—F1 hybrid seeds—was central to this Government GTP agenda of transformation, according to Alemu (Interview with Alemu, 2 December 2013). SP 782/2013 was seen as a seed law, which enabled this push for commercial certification of seed/crops in the formal sector, a sector which despite political changes since the Communist Derg period had remained dominated by the public sector. That was now changing, according to Alemu (Alemu 2011). Feyissa corroborates this point, stating: "The F1 hybrid push is on in Ethiopia now" (Interview with Feyissa, 26 November 2013). This was echoed by Sue Edwards of ISSD: "It is the Government trying to bring subsistence farmers into the monetised economy" (Interview with Edwards, 25 November 2013). However, Feyissa explains the tension surrounding it, stating unequivocally that:

> It is a sovereign security issue for me. Seed is not a commodity in Ethiopia. It is an essential asset that needs to be protected. Do your business but leave the seed alone. (Interview with Feyissa, 26 November 2013)

Feyissa regarded the new law as "a preparation for UPOV 91, as a donor-forced imposition, a market expansion that will create poverty and does not reflect the farmers as it did not include any forum for national consultation",[33] stating that his own position in the first drafting was to

---

[33] None of the domestic NGO representatives interviewed, who work in this area of agro-biodiversity conservation, sustainability and participatory plant-breeding methods or farmers' rights, that is Movement for Ecological Learning and Community Action (MELCA), ISSD, ABN and AFSA, were consulted in formulating 782/2013, which for these groups remains a commercial seed law aimed at commodification of the seed sector. The exemption allowing seed saving and selling in Article 3.2 is the only feature of seed sovereignty that these groups could be satisfied with arising from 782/2013. This is significantly more than other African countries have been able to hold on to in seed legislative changes at this time, notably Kenya's SPVAA 2012.

question "why it was needed to change the strategy at all—asking what is going to be transformed?" (ibid.).

Alemu, on the other hand, saw the big challenge to this opening up and liberalisation of the seed sector to private development precisely in this "centralised state control" (Interview with Alemu, 2 December 2013). Nevertheless he pointed out that the government "have a genuine argument and fear that if we just liberalise today without any capacity the whole market will fail. So you know they have a genuine reason for holding it" (ibid.), meaning holding back full liberalisation. This fear, which essentially is a well-based fear of famine, had cost the two previous regimes their hold on power. It explains the ambivalence of the MoA and Prime Minister's office in relation to advancing 782/2013 throughout the process.

Clearly, the domestic actors could not be entirely ignored in the Ethiopian case, not least by the EPRDF Government. Some of their experts were deeply embedded within government circles and institutions, (even without any political affiliation), and included globally acclaimed Ethiopian scientists, some of whose innovative work on genetic resources at MoA and elsewhere was highly regarded nationally and internationally, within and outside the party. On top of that and notwithstanding the authoritarian nature of the Ethiopian version of the 'DS', subsistence farmers provided the electoral base of the Tigrayan-dominated EPRDF party machine. Those that spoke for them had the ear of government at least to some extent, or at least needed to be seen to be appeased. MoA Director of Regulation, Daniel Mekonnen, pointed out that "the priority agenda of the government in the case of 782/2013 was to protect the farmers, explaining the insertion of the exemption for farmers' varieties", observing the difference with other countries where the law only grants limited access (Interview with Mekonnen, 11 November 2015). This explains to some extent the consistent inclusion of farmers' crops and free community exchange of Plant Genetic Resources (PGRs), now enshrined in the exemption in Article 3.2 of 782/2013 and the insistence of a differentiated approach through acknowledging the multiple seed system pathways within the country. Feyissa acknowledged that "the inclusion of the exemption in Article 3.2 is good, because it states clearly that this Proclamation may not be applicable to the use of farm-saved seed 'by any person'" (782/2013 Article 3.2 a). "So UPOV 78 or UPOV 91 are not here in this exemption which is good, because UPOV 78 and UPOV 91 qualifies that farmers cannot sell beyond a certain volume" (Interview with Feyissa, 10 November

2015). However, Feyissa is still critical of other aspects of the new seed law. He states that "782/2013 in fact removed that space which had been given to farmers in 206/2000 for varietal development, it removed farmers' role and farmers' rights in that and the planned genetic resource base is not there" (Interview with Feyissa, 26 November 2013). For Feyissa, the question always was: "How can we be an agricultural country when we do not recognise farmers as seed producers, when over 96% of the planting material is produced by our small holder farmers?" (ibid.).

One major reason lay in the deepening influence of the seed agribusiness and other powerful external actors/donors who were increasingly embedded with the EPRDF-run government at many levels. These actors could not be ignored now either. The extent and depth of that penetration became more evident as drafting continued. In the meantime the government were intent on ensuring that they controlled and orchestrated as much of that change as possible, at all times maximising the power of the executive and ensuring their domination at all levels of decision-making throughout the drafting process. This occurred even despite the death of the Prime Minister Meles Zenawi in August 2012, before the Seed Proclamation was completed.

The significant, and some would say successful, transition to the commercialisation agenda was accompanied by this intensification of external actor involvement/investment and influence over seed/agricultural policy in Ethiopia (Alemu 2011; IDS Bulletin 2011) during these crucial years from 2008 until the passage of the seed law. This was particularly evident in the government's binding connection to the World Bank's AGP together with United States Agency for International Development (USAID)'s central role in the seed and marketing aspect of that programme, namely, Agricultural Growth Programme (AGP)-Agribusiness and Market Development (AMDe), as well as their collaboration with Du Pont Pioneer's activities under the G8 North American Food Systems Network (NAFSN)/Grow Africa platform, which I shall return to later. It became specifically evident in the major push for changing seed regulations from key actors IDLO, Dutch state interests and later in the drafting process with ATA and the BMGF. It was also central to the intensifying push regionally from within East African Community (EAC), through ASARECA and COMESA for harmonisation of seed laws and regulations.

In this regard, two directors at the MoA, the Director of Inputs and the Director of Regulation, both qualify Feyissa's earlier point regarding the forces pushing for globalising rules to apply. Daniel Mekonnen, as Director of Regulation and Chair of the Technical Committee at MoA, which drafted the new Seed Proclamation, identified the main strands that were brought forward through 782/2013. He cites the limitations of 206/2000 and the need for alignment/harmonisation with regional IP PBR Agreement COMESA as the key drivers determining the core changes in the revision of Ethiopia's seed law. "COMESA harmonisation is now happening in 19 countries. They are organised in three main areas (1) Phytosanitary—protects export, (2) Seed Certification and (3) Variety Release" (Interview with Mekonnen, 11 November 2015). These are key constituent parts of a new strict global IP architecture controlling commercial/trade agendas, and new to Ethiopia's heretofore tightly state-controlled seed sector. "We have to harmonise our standards, therefore in the future seed will easily go from one country to another" (ibid.). This is borne out by key documentation from ASARECA and COMESA, which confirmed that Ethiopian state actors were involved since 2001,[34] working on five key areas agreed for harmonisation: (i) variety evaluation, release and registration process; (ii) seed certification process; (iii) phytosanitary measures; (iv) plant variety protection; and (v) import/export documentation (Waithaka et al. 2011, p. 8). The Director of Inputs at MoA echoed this need for alignment with COMESA as a key motivator for change, whilst also citing three other crucial aspects which were sought and satisfied under 782/2013:

> To encourage local private companies to satisfy seed demand
> To engage seed co-operative unions in seed production
> To improve our quality system. (Interview with Teshome, 11 November 2015)

Mekonnen highlighted that in the second phase of drafting, from November 2009 to September 2010, the Dutch Government and INGO IDLO brought an international/legal influence and opened up the process for a time to other stakeholders (mainly private sector, horticultural exporters). This explains how many of the key features of 782/2013 were

---

[34] This began as a 'Seed Initiative of the Eastern and Central Africa Programme for Agricultural Policy Analysis' (ECAPAPA) which initiated a pilot phase in Kenya, Uganda and Tanzania in 1999. Nine more countries including Ethiopia were added in 2001 and became part of the Seed Regional Working Group (SWG), which in 2004 became EASCOM, whose core mandate was to review seed laws, policies and regulations with a view to regional harmonisation of seed (Waithaka et al. 2011, p. 8).

included, and how some of the concerns of domestic actors thereafter were side-lined, highlighting the role and influence of this key external state actor and an influential IGO at a critical moment in the drafting process.

IDLO's stated task in relation to Ethiopia's 782/2013 law had a strict commercial and legal emphasis as they sought to "develop a new regulatory framework" based on the premise that:

> the legal and institutional framework for agriculture had become outdated and largely ineffective, in particular with respect to international standards and practices. As a consequence, weak quality controls allowed substandard seed to be put on the market, burdensome testing requirements allowed or prevented new varieties from becoming available and inadequate incentives and protections discouraged private investment in the seed sector. (IDLO 2013)

These are the central tenets of globalising seed law, as enshrined in COMESA, ASARECA and UPOV objectives in changing domestic seed laws, albeit with the consent of the Ethiopian Government and key officials in the Ministry of Agriculture at that time. The case for legislative change had also been heightened by export difficulties arising for the Ethiopian flower business.[35] IDLO were also engaged in other areas of seed legislation and cite their involvement in carrying out 'training and technical assistance programmes' in Ethiopia and other African countries at this time, as well as being critically involved in "IP activities geared toward enhancing the ability of government officials and private sector stakeholders to apply and enforce IP laws and improve IP legislative framework".[36]

Despite the contradictory evidence as to who approached whom in relation to the ten-month (November 2009–September 2010) IDLO project, two more drafts (3 and 4) of the Seed Proclamation were produced collaboratively by IDLO and MoARD (Interview with Daniel MeKonnen 11 November 2015; IDLO 2013), as well as some assistance from the Dutch Embassy. The Dutch role is important. Hassena et al. (2016) high-

---

[35] Some export problems emerged when an Ethiopian grower had his 20,000 roses confiscated at Schiphol Airport in November 2008 and was charged with illegally selling them in the EU, which according to Hassena et al. (2016), arose because Ethiopia had failed to enact IP PVP legislation and consolidate proper export arrangements.

[36] http://www.oecd.org/aidfortrade/47027163.pdf [accessed online 4 October 2016].

light that the Dutch Embassy 'works with the horticultural sector', signalling Dutch state interests in the Ethiopian context. However, the Dutch Government also played a role in bringing forward the differentiated seed system approach. They lobbied internationally at this time on behalf of LDCs to "be given scope if they so desired to make use of the flexibility provided under the TRIPs agreement" against the uniformity imposed by UPOV 91 and in favour of participatory breeding programmes (Minister Bleker cited in De Jonge 2014). Even so, Hassena et al. (2016) assert that when a workshop was organised in March 2010 which attempted to bring forward the wishes of the horticultural stakeholders, which included the establishment of an office for seed sector governance, including an exemption for specific varieties from registration to facilitate international trade, it was not warmly received by MoA. It became apparent that "there was no room for the APHRD to negotiate" within the process (Hassena et al. 2016, p. 89). Though the export issue was later dealt with and included at (Part 4, Article 17.2), seed sector governance would not be separated from Ministry/Government hands, despite external Dutch/horticultural industry efforts and would not be included in the final Seed Proclamation, a distinct feature of 782/2013.

Hassena et al. (2016) assert that the whole project (now draft 4) was put on hold, largely because the Ministry did not submit the draft to the PM's office for a year, citing the elections of May 2010 and changes in ministerial positions as reasons for the delay. But they also state that by the time a revised draft 4 resurfaced in October 2011, the Dutch Embassy had not been consulted and henceforward were excluded from the drafting process. However, two other reasons are more likely considering later outcomes. These relate directly to choices made by the EPRDF-run Government and who they chose to align with at this critical juncture. Firstly the Dutch appear to have fallen foul of their hosts at a crucial stage, both in relation to the misappropriation of Ethiopian Teff (an important indigenous, nutritious grain) by a Dutch company, which went bankrupt in 2009 but which nevertheless proceeded through the formation of new companies to claim patent royalties.[37] A residue of disdain for the Dutch remains. Feyissa states that:

---

[37] Teff varieties the Dutch company had taken arising from an MoU signed with EBI which has to date not been resolved and has never benefitted the Ethiopian State and its communities.

the Dutch were the first to hijack the access and benefit-sharing agreement of CBD with regard to Teff, by patenting what they had taken from Ethiopia. Now they are promoting farmers as seed producers but coming in the backdoor to change our seed laws. (Interview with Feyissa, 26 November 2013)

Secondly, the incident at Schiphol airport which criminalised the Ethiopian business was seen as a strong-arm tactic of the Dutch to force globalising Intellectual Property plant variety protection (IP PVP) rules on Ethiopia.

More likely, the Ethiopian State chose to realise its agricultural and seed 'transformation agenda' with a different set of collaborators, namely key US interests, a sanguine political decision based on realpolitik. This was greatly facilitated by the arrival of the BMGF, which had first established contact with Meles Zenawi in 2008. By 2010 it was centre stage, right at the heart of Ethiopia's Government and decision-making on seed futures, while other significant US actors were positioned to dovetail with this entirely new US-dominated programme, which all other actors could come behind.

Significantly, just as the World Bank was constructing AGP as its new funding platform for Ethiopia in 2009, the Gates Foundation had its alleged 'fortuitous meeting' with Meles Zenawi, which led quite promptly to the establishment of ATA by Federal Regulation in December 2010. This heralded the emergence of another 'peculiar' feature of the Ethiopian seed arena, where a new state organisation with core funding from a private American philanthrocapitalist with strong links to seed TNCs (namely Monsanto), alongside other donors, would direct the agricultural changes of the country for the next two decades, starting with the seed law. This provided the prerequisite legislative framework for dismantling the public seed programme and liberalising and privatising the seed sector.

The ATA would be directed by a Transformation Council, packed with EPRDF Government Ministers and key Gates Foundation personnel and headed by the Prime Minister of Ethiopia. The ATA state that:

we wanted to draft a new law because we wanted to make seed into a business, not a service as it has been administered by the Government to the community. We want it to be a business which involves many players. (Interview with ATA representative, 20 November 2015)

ATA state that their role in the drafting of 782/2013 was to

> create a platform to discuss the development of policies, of laws and strate-
> gies. ... ATA organised a technical team to work on each Article and para-
> graph and then we organised a series of workshops with other key partners,
> such as EIAR, MoA to discuss the details. These meetings happened on a
> weekly basis or probably bi-monthly meetings. (Interview with ATA repre-
> sentative, 20 November 2015)

Yit Barek of ATA is now the head of Seed Law Development at ATA,
but was a participant in these workshops at the time as a private-sector
seed producer of hybrid maize. "That is the key driver here", according to
Barek. "Without hybrids the public cannot survive" (Interview with
Barek, 20 November 2015). He pointed to the 'significant differences'
between 2000 law and the 2013 one, stating that 782/2013:

> relaxes the pressures of producers. It relaxes in relation to new varieties. It
> relaxes the source of varieties and it gives power to the regulatory structure,
> increases capacity to check quality from field inspection right down to the
> seed selling point. (Interview with Barek, 20 November 2015)

Barek saw ATA as a 'facilitator and harmoniser' and confirmed that the
ATA final drafting process phase of 782/2013 was very much an Ethiopian-
dominated process. He stated that:

> There were firstly regional facilitations, then national ones, which took place
> in big Assembly halls, not at ATA offices (ibid.). At first, the public and
> private seed producers were there as well as the Ethiopian Seed Enterprise
> (ESE), and also the regulatory, extension and marketing people of MoA and
> BoARD as well as EIAR, who were the major stakeholders. Then when the
> drafting was done, the legal sector in MoA reads every line, and that unit
> reads it to see it is in accordance with environment policy or whatever and
> once that unit is comfortable they send it to Parliament, and from there fol-
> lowing Parliament approval to be signed by the President. (ibid.)

Of significance was Barek's indication of support for the exemption for
smallholder farmers and farmers' rights, stating:

> You cannot regulate that ... the law is for the commercial transaction which
> must be regulated, but represents only a small percentage of the population.

Whereas the farmer-to-farmer exchange, you cannot regulate that ... you have also to respect farmers' rights, the exemption is good. It is also for food security. You cannot force everyone to buy seed. (ibid.)

The ATA model clearly appealed more to the Ethiopian Government. It allowed them to hold executive power and supposedly to have substantial input into the unique design of their seed agenda. Simultaneously, it opened the door through key aspects of the law to specific seed developments that would substantially benefit private-sector development, liberalisation and decentralisation of seed business to the regions, all of which served US interests in particular, but could be carefully managed through the government-controlled agencies and agents throughout the country.

In this way the new ATA, a unique hybrid of an authoritarian Ethiopian executive and an unaccountable private US philanthrocapitalist transnational actor used the leverage of the drafting of Ethiopia's seed law between 2011 and 2013 to open the door for other key US actors to enter the seed space and shift seed sovereignty in a profound way. The timing was perfect.

The World Bank launched AGP on 16 February 2011. The US Government through USAID "parallel funds the World Bank AGP but works through the GoE and with the Ministry of Agriculture/ ATA collaboratively and with Regional Government Institutions as well as local" (Interview with USAID representative, 19 November 2015). It now assumed responsibility for the entire sub-component "Market and Agribusiness Development" (AGP-AMDe), and were thereby empowered to make interventions and support the seed sector (World Bank 2010, pp. 12–13).

Regarding the new seed law, USAID confirm that they would have been "part of the enabling environment to allow this to come through, to try and get more private interest in the seed market and to stimulate market/commerce" (Interview with USAID representative, 19 November 2015). This private-sector stimulation brought an immediate change in the practice of seed sovereignty, as USAID confirms that "Du Pont/ Pioneer brings plant material from South Africa and grow it out on private commercial plots to test and then distribute to farmers for marketing, so that it can be sold" (ibid.). This 'purchase programme' is targeted at 35,000 farmers already who are now buying seed from a giant TNC, tied in to World Bank AGP and USAID programme, all in tandem with the ATA/GoE/MoA agenda for seed/agricultural industrialisation (ibid.).

This was confirmed in interview with Daniel Mekonnen, Director at MoA. He stated:

Now, with 782/2013 in place, TNC seed companies such as Du Pont/ Pioneer can play a much greater role in the country, because if the farmer has enough money to buy good quality seed, he likes it. (Interview with Mekonnen, 11 November 2015)

For USAID, which authorises Du Pont to run their seed programme in Ethiopia, the underlying objective in relation to the quality maize project is to "try to shift from Open Pollinated Varieties (OPVs) to hybrid maize for productivity" (Interview with USAID representative, 19 November 2015). This marks a key shift in seed sovereignty. Of interest is that USAID has been working through a public-private partnership (PPP) since 2012 between the Ministry of Agriculture, ATA and Du Pont Pioneer, trying to popularise hybrid maize seed in three regions Amhara, Oromia and Southern Nation (Ibid.), acknowledging that it is "taking a while for farmers to let go of their own varieties, especially regarding taste, but farmers are adapting" (ibid.). In other words, leaving their previous seed practices behind. The new Seed Proclamation 782/2013 enabled these fundamental changes in seed sovereignty to occur.

CONCLUSIONS FROM THE ETHIOPIAN SEED LAW CASE

Consistent with transformationalist accounts, the new Ethiopian seed law ambiguously gave recognition to a three-tiered highly differentiated seed system—a reflection of the myriad actors involved in the process and the influence and authority they exercised throughout. This law created a differentiated standardisation and certification system for certified improved seed, separate to a less stringent system for quality declared seed, whilst also including a complete exemption for the 97% of the farming population who rely at present on the informal non-commodified seed system. The exemption is significant and unexpected in the present tranche of seed legislation occurring on the continent, where pro-PBR/PVP invocation of the farmers' privilege is outlawing seed saving and sharing amongst subsistence farmers and their seed systems. However, despite this clear recognition of farmers' seed systems, the government chose to simultaneously adhere to globalising UPOV 91 rules in certain sections of the law, though not required as an LDC country, or under TRIPS.

Yet the process of law making reveals highly complex and contradictory features, consistent with the transformationalist model. From the early stages of drafting of 782/2013, it became apparent that the opinion and influence of experts at the PM's office would direct the passage of the law. Key domestic actors were included in the Ethiopian case, unlike the Kenyan legal process for SPVAA 2012, notably ESE, EIAR, EBI and ESA. However, despite a degree of consultation, some of these organisations themselves expressed some ambivalence with regard to their influence and the subsequent outcome. For example, though the farmer's exemption, which EBI would have sought, did persist throughout the process, all mention of EBI itself was dropped from the final text, coinciding with the ATA presiding over the drafting process from the fourth draft. This signals a reduction in influence in the seed space for domestic actors, particularly those with a pro-farmers' rights/ participatory seed sovereignty viewpoint. Furthermore, no civil society organisations (CSOs) were consulted at any stage of drafting. The EPRDF-led Ethiopian Government, as the key state actor, retained dominant control over seed governance in both the formal and informal seed sector. However, despite their dominant state position, they were weakened considerably by the increasing intensification of other actor's involvement in the seed space, which the drafting process revealed.

The new Seed Proclamation announced an unprecedented opening up to liberalisation indicating a certain disaggregation of state control and function in the seed space, not least because of the presence and influence of a number of key external actors, notably the World Bank, BMGF, USAID, IDLO and the Government of the Netherlands. These actors were highly influential, at various stages of the drafting process. The manner in which the Ethiopian state shifted loyalties between these actors in attempting to bolster their own strategic advantage in a changing landscape in seed sovereignty is entirely consistent with the contingent, contradictory accounts which transformationalist scholars assert is the hallmark of globalisation in the new polycentric disaggregated political space.

This case reveals the key role played at a critical stage of the making of the seed law by the BMGF. This provides a unique insight into how new supranational transnational actors (TNAs) such as these interact with domestic governments, affecting profound changes in the public policy space. Their role can best be understood with reference to the transformationalist school of thought. It contains all the ambiguities and contradictions which transformationalists refer to. BMGF was a powerful new actor.

It became a central player—initially through funding the diagnostic studies which became the building blocks of their ultimate goal—to establish an elite ATA, deeply connected to the executive wing of the government. This ATA proceeded to operate thereafter without the tedium of parliamentary accountability and bureaucracy. This proved efficient and successful in bringing the new Seed Proclamation to a conclusion rapidly. BMGF also had crucial power and influence over other actors. It was a bridge between an authoritarian state and already dominant US interests, notably US Government, AGRA, International Centre for Tropical Agriculture/ Consultative Group on International Agricultural Research (CIAT/ CGIAR), IDLO, World Bank and key seed TNCs. It had global leverage. It received domestic support. Its diagnostic studies prepared the ground for seed legislative changes to come. It established a new agricultural agency which deepened the ambiguity, by merging a giant private philanthrocapitalist foundation with the PM of an authoritarian state, including the executive of government on a transformation council which effectively would make decisions on such profoundly important matters, outside of what was already an undemocratic parliament. This sounds hyperglobalist, except that the Ethiopian law opted for a differentiated seed system and an exemption for smallholder farmers and pastoralists. Seed sovereignty was maintained in the hands of majority smallholder farmers. ATA specifically stated they could not be legislated for, because of farmers' rights and needs. It embodies all of the characteristics of transformationalist thinking in this regard.

In the critical period following their own disputed landslide electoral victory in 2010, the Ethiopian Government finalised the seed law, choosing to embark on an almost exclusive collaboration with US-dominated interests. Simultaneously, they adroitly orchestrated the process to ensure a differentiated seed system would be enshrined in the Ethiopian law. The EPRDF-dominated state remained a central player, but a disaggregated one, as they adroitly managed their own exit from certain areas of sovereign seed control, conceding power and relinquishing responsibility to a private transnational actor (BMGF), its key geopolitical/military ally, namely the USA, and main IO funder, the World Bank. Nevertheless, certain alternative and pluralistic outcomes in terms of seed sovereignty were realised in this case, albeit by an authoritarian government in a less than transparent fashion and without full public consultation. This is entirely consistent with transformationalist interpretations of globalisation.

REFERENCES

ACB. 2015. *The expansion of the commercial seed sector in sub-Saharan Africa: major players, key issues and trends* [Online]. Available from: www.acbio.org.za [Accessed 3rd February 2016].

Alemu, D. 2011. The political economy of Ethiopian cereal seed systems: state control, market liberalisation and decentralisation. *IDS Bulletin*, 42(4), pp. 69–77.

Asmelash, H. 2014. On Ethiopia's Long Walk to WTO Membership, U. of St. Gallen Law & Economics Working Paper No. 2014–06.

ATA. 2015. *Agricultural transformation agenda progress report covering 2011–2015 in the GTP I period*. Addis Ababa, Ethiopia: ATA.

Atilaw, A. 2010. *A baseline survey on the Ethiopian seed sector*. Addis Ababa, Ethiopia: African Trade Association.

Bioversity International. 2015. *National implementation of the international treaty on plant genetic resources for food and agriculture* [Online]. Available from: http://treatylearningmodule.bioversityinternational.org/fileadmin/templates/learning/upload/eng.policy_module/Handouts_1-12/Handout%20 9%20National%20implementation.pdf [Accessed 10th April 2017].

Broek, J.A. van den. 2015. Business opportunities report seed #4. *Centre for Development Innovation* (CDI) [Online]. Available from: https://www.rvo. nl/sites/default/files/2015/11/Rapport_Seed_Ethiopi%C3%AB.pdf [Accessed 10th April 2017].

De Jonge, B. 2014. Plant variety protection in sub-Saharan Africa: balancing commercial and smallholder farmers' interests. *Journal of Politics and Law*, 7 (3), pp. 100–111.

Feyissa, R. 2006. *Farmers' rights in Ethiopia: a case study*. Lysaker, Norway: The Fridtjof Nansen Institute.

GRAIN. 2008. Seed aid, agribusiness and the food crisis. Editorial. *Seedling*, October: 2–9.

Hassena, M., Hospes, O. and De Jonge, B. 2016. Reconstructing policy decision-making in the Ethiopian seed sector: actors and arenas influencing policymaking process. *Public Policy and Administration Research, 6(2), pp. 84–95*.

IDLO. 2013. *Strengthening Ethiopia's agricultural sector* [Online]. Available from: www.idlo.int/english/WhatWeDo/Programs/Ethiopia/Pages/ProjectsDetails. aspx?IDPRJ=61 [Accessed 2nd November 2013].

IDS Bulletin. 2011. *The politics of seed in Africa's green revolution*. Oxford: Wiley and Blackwell.

Kloppenburg, J., 2014. Re-purposing the master's tools: the open source seed initiative and the struggle for seed sovereignty. *Journal of Peasant Studies, 41(6), pp. 1225–1246*.

Louwaars, N.P., De Boef, W.S. and Edeme, J. 2013. Integrated seed sector development in Africa: A basis for seed policy and law. *Journal of Crop Improvement.* 27, pp. 186–214.

Munyi, P. and De Jonge, B. 2015. Seed systems support in Kenya: consideration for an integrated seed sector development approach. *Journal of Sustainable Development,* 8(2), pp. 161.

Munyi, P., De Jonge, B. and Visser, B. 2016. Opportunities and threats to harmonisation of plant breeders' rights in Africa: ARIPO and SADC. *African Journal of International and Comparative Law,* 24(1), pp. 86–104.

Waithaka, M., Nzuma, J., Kyotalimye, M. and Nyachae, O. 2011. *Impacts of an improved seed policy environment in Eastern and Central Africa* [Online]. Available from: https://www.asareca.org/~asareca/sites/default/files/publications/Impactsofanimprovedseedpolicyenvironment.pdf. [Accessed 2nd June 2016].

World Bank. 2010. *Project appraisal document on a proposed credit and a proposed IDA grant to the Federal Democratic Republic of Ethiopia for an agricultural growth project.* Sustainable Development Department, Agriculture and Rural Development Unit, Country Department 3, Africa Region: World Bank.

CHAPTER 6

The Ethiopian Highlands: The Exercise of Seed Sovereignty at the Local Level

This case study brings the examination of seed sovereignty down to the local level to one of the nine autonomous regions of Ethiopia, the highland region of Oromia. I have chosen to look at the seed practices that have emerged in the Arsi locality in this region. Here, some farmers have started to plant externally sourced malt barley seed for the corporate-dominated beer industry on a sizeable portion of their once traditional subsistence plots of land. However, in the same locality farmers are also maintaining and enhancing landrace barley varieties (i.e. farmer-selected, produced and locally adapted seed variations) (Alemayehu 1995). These farmers' varieties are synonymous with the participatory, informal seed sector associated with what we define as seed sovereignty as outlined in Chap. 3.

This is an important case study. It captures a change in land use with the introduction of contractual seed from an external corporate seed source from outside the local agroecological setting. This might signal a loss of seed sovereignty by local farmers. At the same time, those farmers are continuing to plant their own locally adapted seeds, suggesting that they still exercise seed sovereignty in this regard. This case examines what is happening in this locality with regard to these changing seed practices. Who decided to make the change to using contractual seed? How did they decide? Who was included in the decision? Who was excluded? When did it happen and how?

© The Author(s) 2019   197
C. O'Grady Walshe, *Globalisation and Seed Sovereignty in
Sub-Saharan Africa*, International Political Economy Series,
https://doi.org/10.1007/978-3-030-12870-8_6

These questions were posed to all actors, whether CEO of Diageo, Self Help Africa (SHA) representative, ATA personnel, Ministry of Agriculture official or to the farmers in the field. This chapter presents their views and analyses them in the context of the differing approaches to globalisation. Which approach best captures the changes that are happening in seed practice? Hyperglobalists would say that corporate seeds and value chains are taking over where state/public system once led. Sceptics would see the state as the retaining its role as the main fulcrum of power, despite certain changes. Transformationalists would expect to find a contingent, differentiated seed system, replete with new contradictory but overlapping sovereignties, in new shared polycentric arrangements of seed governance.

I divide this chapter into three sections. The first section presents the two types of seed practice in this barley-growing locality. I present the recently introduced seed practice where farmers are engaged in the production of malt barley seed and grain for the corporate drinks industry in 'clustered value chains' on their now 'pooled land'. I then describe the other seed practice in the same locality which relies on own-saved farmer-selected varieties, where predominantly food barley is grown, alongside other subsistence crops in 'segregated plots'. I set out the key constituent components of their seed practice, highlighting the different choices they are making on-farm and in the cooperative community. The next section examines the processes by which decisions were made that resulted in such a differentiated outcome on the ground. The final section focuses on the actors involved in determining the differentiated choices and seeks to explain their motivations in this case.

Seed Practice in Oromia Region

There are nine autonomous regions within Ethiopia. Oromia is one of them. Oromia is the largest regional state in terms of size and population with 27 million people, one-third of the total population of the country. It also experiences the highest rainfall and thus the highest level of cultivation (Ethiopian Census 2007) and has long been targeted for cereal intensification (Spielman et al. 2010, p. 187).

In the Arsi region of Oromia, I focus on the Primary Cooperative (PC), 'Senaboru', which is part of a larger Gallema Farmer's Cooperative Union. This PC is located within the kebele/neighbourhood peasant association (PA), "Digeru Bora", which is within the worede/district of "Digeluna

Tijo". The Digeru Bora PA consists of 107 members, 100 male and 7 female (Interview with Self Help Africa (SHA) Agronomist/field officer for Peasant Association (PA), 'Digeru Bora', 24 November 2015).

Historically, the farmers in this area allocated almost their entire plot to the production of food barley to fulfil their subsistence needs. Even those who were involved in more intensive production were not organised into commercially focused, functioning, cooperative union structures according to SHA. Some informal cooperatives existed there from earlier times and any surplus produced was sold in local markets, with farmers generally bringing their "barley to markets 5–20 km away from their villages, using pack animals" (Kaso and Guben 2015, p. 91). Some farmers grew malt barley for the nearby state-owned Assela brewery, but "high climate and price risks reduced incentives to intensify" (Spielman et al. 2010, p. 191), so most production centred on food barley. Thus, the local beer factory relied on imports from abroad.

However, in recent years farmers in the locality have started to use an 'improved seed', an Open-Pollinated Variety (OPV) of malt barley called 'Travlr'. This is an externally derived seed that is used for commercial purposes (Interview with Self Help Africa (SHA) Agronomist/field officer for Peasant Association (PA), 'Digeru Bora' 24 November 2015). It is being supplied to the farmers by an INGO, Self Help Africa. One farmer described the situation as follows:

> SHA brought 60 Quintals of 'Travlr' Variety malt barley seed to our PC, which are very good and now we have been producing grain for sale to the Assela Factory and at 1,150 Birr we are getting a high profit from this. They provide new seeds, new pesticides, new insecticides. All of these are non-repayable, left for the farmers as an incentive to produce more, to engage more and to be a kind of model for others, and that is why more people are now asking to become members. (Interview at PC, Farmer Tasfaye, 24 November 2015)

SHA explain that these farmers, who normally would have "on average one hectare of land, have pooled some of their land resource and formed a 'cluster', which is technically important for pest control and seed production" (Interview with SHA Manager Tadenya, 20 November 2015). The farmers explain how this is done and showed me the barley fields where they have carried out this adjustment:

This is achieved through larger, uniform clustered/pooled land areas, where improved malt barley is grown without other varieties, such as food barley. We are producing nowadays for the market. (Interview with Farmers at PC, Gallema Farmers' Union, Arsi, 24 November 2015)

The farmers explain that this 'cluster' makes it easier for them to "access different technologies and get information about orienting ourselves for the market" (Interview at PC, Farmer Tasfaye, 24 November 2015).

This change of practice is assisted by SHA's encouragement of farmers to grow malt barley. SHA points to new agreements with drinks companies, Diageo in this case, that allow farmers to get a better fixed price for their malt barley produce, providing them with greater market certainty, a key element of ensuring this switch from the "informal to the commercial formal seed system" (Interview with SHA Manager, 20 November 2015). There is now a contractual market agreement between the farmers and Diageo to produce malt barley and the price is set through the signing of a Memorandum of Understanding between the farmer cooperative as well as the larger Gallema Farmer's Union and Diageo and is thereafter included in the byelaws of the cooperative. In the case of Diageo's contractual agreements, Diageo explain:

We pride ourselves on the fact that we really invest time and energy in training farmers, having direct contracts with each farmer and working closely with farmer cooperatives and unions. The package we provide our farmers includes training in the field, interest free pre-financing, quality seeds and fertiliser. (Phone Interview with Geoff Wallis, CEO Diageo Ethiopia, 26 November 2015 and email questionnaire response, 3 December 2015)

Diageo further explain that "this is a binding contract, guaranteeing the farmers access to credit, training, improved seeds, fertiliser, crop insurance, pesticides and an end client for their produce" (ibid.).

From this, it is evident that the pivotal switch which has occurred is represented by the introduction of a new externally sourced seed variety, which is part of a package of a new corporate/contractual seed practice and this has successfully been established in this locality. The newly appointed commercial malt barley farmers of this PC and in the wider Gallema Farmers' Union Cooperative in Arsi zone describe how they are now using improved, externally sourced, malt barley seed, the majority to produce grain for beer production and some designated to produce malt barley seed from the new varieties. They show how they are pooling their seed sovereignty in allocating their now conjoined land through their PC, to a new commercially oriented seed 'cluster' for commercial seed development for the corporate beer industry, and improving their livelihood in various ways in doing so. They state that "with increased yield and productivity and obviously increased quality then the price improves and with that the life of the individual farmer" (Interview with Farmers' Cooperative, 24 November 2015). SHA Manager confirms that the Government Certification process follows and "in this way these farmers have changed from the informal to the commercial formal seed system. That is the process. In Arsi they are now producing for the formal seed system" (Interview with SHA Manager, 20 November 2015). This switch from planting locally adapted seeds for food barley for subsistence production to externally sourced seeds for commercial malt barley production suggests a loss of seed sovereignty in this locality.

However, extended interviews with the farmers revealed another aspect to the story. They explain:

> Now that the land we are producing on is decreased we are limited because we have to use a bigger part of the land for the 'improved seed', because there is some value attached to it. (Interview with farmers at PC Senaboru, 24 November 2015)

This quote hinted at the fact that they have not pooled all of their land and that they are not planting only corporate improved seed. When asked, they confirmed that they maintain their own on-farm 'segregated plots', which are dominated by traditional food barley varieties. These plots are away from the commercial malt barley seed to avoid cross-pollination. In other words, they are continuing to use their own traditional seed practice for their household/community needs. Indeed, evidence suggested that only approximately half of the total land resource of each farmer in this locality had been pooled to benefit production for the SHA/Diageo malt barley cluster there (Interview with Adam Sano, Government Development Agent (DA) at Senaboru PC, 24 November 2015).

This is puzzling. With such global forces involved I expected to find evidence of extreme hyperglobalism, with an entirely commercially oriented, corporate-dominated, externally sourced malt barley production in this locality and within this cooperative, as it is a key area targeted for malt barley clustered value-chain development. Yet it becomes evident that the vast majority of the farmers in this same locality continue to maintain their traditional seed production practice. Dr. Melaku Worede explains this traditional seed system:

> The seed system used in most traditional farming systems is based on the local production of seeds by the farmers themselves. Farmers consistently retain seed as a security measure to provide a back-up in case of crop failures. They always store seeds for three main purposes: consumption; sale; and seed stock (for sowing in the next season). (Worede 2011, p. 368)

Worede's explanation is informative. He outlines the importance of farmer's practice of seed selection, production and saving for 'informal distribution of planting materials within and among the farming communities'. Seed production in this system is largely "non-specialised; an integrated production of field crops for consumption and marketing, and depends almost entirely on 'famers' varieties' with the exception of cases where the seed system depends on improved or introduced crop varieties" (ibid.).

This includes growing a wide variety of indigenous self-pollinating barley varieties, each with different socio-cultural and household uses (including predominantly food and some malt barley varieties for household/local consumption). This allows them to maintain year-round food security because of the different planting and harvesting times, for 'Meher' (main season) and Belg (spring short season). This is in keeping with widely documented research of barley's continued popularity as an early harvested crop and essential 'hunger breaker or relief crop during a season of food shortages' in Ethiopia, particularly in the Highland region (Bayeh and Berhane cited in Kaso and Guben 2015, p. 85).

In this way, these farmers are maintaining broad in-situ (on-site) heterogeneous seed practices, which they say they "like very much and is very good quality" (Interview with farmers at PC Senaboru, 24 November 2015). The farmers explain that they do not compare the two types of seed practice they are now engaged in. They give "more value to this improved seed one, because it has more value and is more use" to them, but they state that "at the same time we maintain the old indigenous varieties, because it has its own use, as food and as grain" (ibid.). In this way the farmers reveal that they are now maintaining two types of seed production practice in the locality, one for commercial malt barley and one for indigenous food barley varieties.

This is a key finding of this case study. Despite considerable focus from government, TNAs and INGOs to develop contractual, commercial malt barley production systems based on externally sourced seed, the farmers here are maintaining two distinctly different seed practices. This seems to suggest agency and sovereignty in decision-making around seed on behalf of the farmers. In the next section I address the processes by which the decision was made to allow two types of barley production to be pursued in this locality at this time.

HOW THE DECISION TO SWITCH TO MALT BARLEY PRODUCTION WAS REACHED

This section seeks to determine how the decision was made to switch a considerable portion of the farmer's landholding to malt barley production using externally sourced seed, and how a paradigmatically different seed practice, based on farmers' seeds is being pursued in the same locality simultaneously.

The story starts at the national level with the decisive macro-economic and political drive by the EPRDF-run Ethiopian Government towards agricultural industrialisation since the 1990s. This led to the push to develop the malt barley value chain and the inclusion of the transnational corporate beer industry to realise this goal. I examine the chain of decision-making which led to the changed seed practice and examine how this played out at the local level.

Domestic State: Ethiopian Government Policies 1990s to the Present

The development of the malt barley 'value chain' became a central part of the Government of Ethiopia's agenda through the country's Growth and Transformation Plans (GTP), GTP1 and GTP2, in 2010/2011 and a concomitant reduction in interest in food barley varietal development (Abay et al. 2009, 2011) and confirmed in interview with Dawit Alemu, formerly of EIAR (Interview with Alemu, 2 December 2013). This coincided with both the new round of World Bank funding under AGP 2009 and most significantly the formation of the Gates-inspired Ethiopian ATA the same year, when both the policy and the agency combined many of their core objectives.

The GTP is set in the context of the earlier Agriculture Development Led Industrialisation (ADLI) programme, which has been the main policy principle of the EPRDF-run Government to promote national development since their overthrow of the military Derg regime in the 1990s. ADLI, coupled with the more recent conversion to the concept of Democratic Developmentalism (DD), a development paradigm pioneered in academia by Chalmers Johnson, pursued in some emerging Asian countries and articulated by the late Ethiopian Prime Minister, Meles Zenawi in 2006, became the driving force of Ethiopian development in this period. DD rejected both the rent-seeking tendencies of other African countries and the neo-liberal paradigm, which it was claimed had been disastrous for African countries in the 1980s and 1990s. The 'developmental state', it was argued, would have to "delay the process of democratisation in the interests of fast-track economic transformation" (Meles n.d. quoted in Ayenew 2014, p. 4), a state model characterised by what Ayenew calls "economic determinism and political authoritarianism" (ibid., p. 26). Both ADLI and DD set in train an action-oriented practical programme "upheld by strong leadership and shared by every official of Government" (Ohno 2009, p. 2). The programmes, including the value-chain model which followed, emanated from this 'transformational' paradigm.

When the ambitious five-year GTP1 was established in 2010–2015 (GTP2 2015–2020), it did so on the back of these state-led programmes, thus furthering the emergence of central government as the dominant actor in economic policy planning and implementation. Ayenew asserts that the GTP copper fastened the existing policy as it "envisages centralised management of the economy and a strong interventionist state to implement many mega projects" (Ayenew 2014, p. 4). The Ministry of Finance and Economic Development (MoFED), the main executive body in charge of its execution, describe GTP as a "comprehensive formal development plan constructed for both federal government and the regions". It had two broad objectives as outlined by Ayenew (2014): firstly, to achieve middle-income status for the country by 2025[1] and, secondly, "to change the structure of the national economy from one of

---

[1] Middle-income status according to the World Bank (2013) is defined as one with a gross national income (GNI) per capita of around US $1430. By 2011 Ethiopia's GNI stood at US $1105 for population of approx. 85 million (Ayenew 2014, p. 14). The population is now estimated to be 104 million http://www.worldometers.info/world-population/ethiopia-population/ [accessed online 2/6/2017].

being predominantly agriculture-based to industrial and services led" (Ayenew 2014, pp. 13–14). Agricultural industrialisation was seen as central to achieving this transformation, and a convincing case was made to support that direction by Ethiopia's primary funder—the World Bank, which heavily influenced the outcome. Of importance here is that The World Bank through AGP and USAID was already operational in commercialising value-chain market and agribusiness development in 96 'targeted woredas' (districts) before the ATA took over delivery of the malt barley value-chain project. These target zones are recognised as "high potential agriculture woredas of the four regions: Oromia, Amhara, SNNPR and Tigray" (AGP 2015, p. 1) and represent a highly contested pro-corporate strategy of the World Bank (The Oakland Institute 2017). Of significance to the barley case study is the market and agribusiness and market development (AMDe) aspect, which specifically identified key 'value chains' as a priority, barley being one.

The arrival of the BMGF and the establishment of the Gates-inspired ATA advanced the GTP and AGP value-chain strategy considerably. Tefera Derbew, Ethiopia's Minister for Agriculture, explains that:

> utilising the objectives and targets of GTP 1 as its foundation, the Transformation Agenda (of ATA) prioritises the removal of bottlenecks that stand in the way of achieving these national goals while mobilising the critical stakeholders to implement interventions most effectively. (ATA 2015, p. 4)

Thus, the ATA was critically empowered to dovetail with the government programme and charged with developing and realising the goals with key stakeholders. EPRDF still had core executive power as the Prime Minister and leading members of the Ethiopian Council of Ministers dominated the Transformation Council of the new agency. It also dovetailed with long-standing funding programmes, most notably the World Bank and USAID's AGP-AMDe, Ethiopia's principal donor agency and key geopolitical ally respectively. Of particular relevance is USAID's management of the marketing and agribusiness, including seed sector of the World Bank AGP.

In 2010 the ATA became the new centre of power and authority in driving the 'transformational agenda'. In keeping with its stated goals in the Regulation that established it, ATA was thereby empowered to "provide leadership in identifying, designing and effectively implementing solutions to basic hurdles of agricultural development as well as providing 'policy directions and leadership'" (Part Two, Article 5, ATA Regulation 198/2010).

ATA, which was moulded in the Gates-style business model with clear results-based goals and time-lined objectives, set out their business plan as follows:

> The objectives encompass: achieving a sustainable increase in agricultural productivity and production, accelerating agricultural commercialisation and agro-industrial development, including value chains, reducing degradation and improving productivity of natural resources. (Agriculture Sector Policy and Investment Framework [PIF], 2010/2011–2019/20 quoted in ATA, p. 8)

ATA identified 84 key 'deliverables' during GTP1, which would catalyse the transformation of the agricultural sector, immediately addressing the key bottlenecks.

"Deliverable 57 (Expand Domestic Sourcing of Malt Barley)" (ATA 2015, p. 69) was identified as a key target that brings significant 'value-added' export possibilities and a reduction in import costs for a burgeoning industry looking to source local barley for production purposes (ibid., p. 68). Thus, expanding domestic sourcing of malt barley became a key component of the ATA agenda.

ATA was empowered to identify and remove any 'systemic bottlenecks' that constrained this barley sub-sector and immediately established it as a key area requiring a "transformational strategy for development of the commodity" (ibid.). To this end, they immediately set about creating an 'enabling environment', through the formation of Agricultural Commercialisation Clusters (ACC) in targeted 'high potential' agricultural zones, which would in turn open the door to private investors and a more liberalised market regime.

Key personnel[2] were seconded to ATA from selected organisations to actualise these decisions on the ground in places like Arsi. Non-profit

---

[2] Such personnel had already been engaged for the preceding years with ASARECA, CAADP and other USAID-funded regional initiatives aimed at harmonising seed laws and opening up economies to liberalisation of the private-sector markets for key seed and crop commodities, which vastly benefit private breeders and bigger corporations over the public sector.

organisation, Synergos,[3] thereafter led the design and implementation of the ACC initiative in Ethiopia throughout this crucial period.

The ACC marked the moment when the privatisation agenda moved centre stage within Ethiopia. A strategist of the ACC initiative describes his role:

> I led and actively engaged in the setting of criteria for the identification of priority commodities and geographic areas based on which 31 agricultural commercialisation clusters were identified in the four big agricultural potential regions of Ethiopia. (Techana Adugna, ACC Lead 2016)

In this way, the ACC Initiative or 'cluster approach' became a central building block of a 'strategic value-chain' alliance/platform which followed.[4] Adugna explains that:

> A value chain alliance is a multi-stakeholder platform of value chain actors to discuss and align on critical issues for the development of one value chain, to identify, coordinate and ensure accountability on critical interventions needed to foster backward and forward linkages. (ibid.)

ATA set out its three clear objectives for malt barley as one of its first key Transformation Agenda Deliverables:

1. Increasing production and productivity of malt barley through an integrated set of interventions, including improved inputs use and best agronomic practices;
2. Improving market access to smallholder farmers through the engagement of market actors; and
3. Putting in place the appropriate structures and market-based incentives to achieve national self-sufficiency in malt barley, in order to end imports (ATA 2015, p. 68).

In essence, this point marks the moment when high-level policy shifted from being a descriptive aspiration to being implemented in the local area.

---

[3] Founded by one of the Rockefeller family in 1986, and benefitting from considerable funding from BMGF, Rockefeller and USAID amongst others.

[4] GTP1 Transformation Agenda work relating to 'Systems' is important, which included the crucial and connected areas of seed, soil, cooperatives and market transformational agendas (ATA Report 2013).

This value-added approach pursued at a federal level and actualised at a regional and local level brought a myriad of other actors into the decision-making and implementation phase on the ground. As identified by Kaso and Guben (2015):

> A value chain is made up of a series of actors (or stakeholders) from input suppliers, producers and processors, to exporters and buyers engaged in the activities required to bring agricultural product from its conception to end use. (Kaso and Guben 2015, p. 94)

In fact, choosing to bring Ethiopian agriculture in the direction of this commercial value-chain approach brought in a whole other layer of technical, business and financial service providers (AGP 2017 [accessed online 4/1/17]) and was a foundational principle of the model which ensued. ATA became the primary broker in establishing the Public-Private Partnership (PPP) approach which emerged, enabling all the key actors, notably, INGOs (SHA), TNCs (Diageo) and the farmers of Senaboru PC and Gallema Farmers' Union to activate the malt barley value chain.

Local Application

Rolling out the 'cluster approach' in a key location in order to realise the federal policy at a local level was imperative at this juncture. Choosing the barley belt communities of Arsi in the Oromia Region was an important decision and determined the seed interventions and changed seed practices, which followed.

ATA state that:

> Arsi was chosen firstly because it has the right agroecology and the right geography for malt barley quality demands. Secondly farmer's skill and knowledge is critical. They are barley farmers. You have to upskill them. You have to have the regimen that farmers produce malt barley and thirdly there is a malt barley factory already based in Arsi ... so already a culture of producing malt barley. (Interview with Yit Barek, 20 November 2015)

This was the critical period when SHA[5] joined with Diageo and ATA to connect these local barley farmers with market outlets/linkages. In this

---

[5] This was backed by a government donor (Ireland Aid) through a bilateral aid programme with Ethiopia.

way, SHA, with a history of involvement in assisting Ethiopia's small-holder farmers to 'increase farm productivity', focused on facilitating 'the development of new market opportunities' and 'connecting SHA-supported small barley farmers' (supplying 6500 households with 'improved malt barley seed') with the brewing company. It was also part of the wider economic development project during this crucial 2012 period to date according to Self Help Africa's Ethiopia Report in 2016.[6]

The Ministry of Agriculture confirm that "this 'cluster approach' pioneered by ATA is now operational with Diageo in Arsi" (Interview with Mr. Teshome Inputs Directorate MoA, 11 November 2015). It followed "three years training farmers to use 'improved seed' to fulfil the requirements of the Assela Malt Factory in the nearby town of Assela in Digeluna Tijo Woreda of Arsi Zone" (Interview with SHA Manager, 20 November 2015). This commercial venture requires uniformity and purity for the new seed certification system of the Government Agency for Quality Assurance, an explicit requirement under the new Seed Proclamation 782/2013.

The Assela Malt factory had been a state-owned brewery and processing factory since its establishment in 1984 and has been the sole supplier of malt barley for domestic breweries, with a production capacity of about four million hectolitres. The development of the malt barley chain initiative has significantly altered this monopoly structure, not only affecting the choice of seed practice of the local farmers, but shifting a previously nationalised factory with a low fixed price set by the state to a liberalised market model with increased corporate intervention determining varied outcomes from price to seed practice. With the advent of PPPs and private/PLCs involvement with the former state-controlled Assela factory, a new competitive market has been established and contractual agreements are being made between companies like Diageo and the farmers as well as directly with Primary Cooperatives (PCs). Now a fixed price is set with Diageo or Heineken or equivalent private company and this is included in the byelaws of the cooperative.

> This is why the cooperative structure is being enhanced in Ethiopia. It is a key feature in the Government GTP strategy and the development of the value-chain approach. (Interview with Tadele Bento, Cooperatives Expert and Policy Development, ATA, 16 November 2015)

[6] https://issuu.com/self_help_africa/docs/ethiopia_country_profile__feb_2016 [accessed online 22/12/2016].

SHA were central to the decision-making process throughout. They acted as the broker to establish and formalise the cooperatives, bringing them to a standard that would allow them to become functional for the development of the malt barley value-chain commercial approach. To this end, they organised the farmers into larger cooperative structures, which in turn linked them to a bigger union (in this case the Gallema Farmers' Union, which consists of 14 primary cooperatives),[7] which greatly facilitated the commercialisation process and the switch to the formal seed system (Interview with SHA Manager, 20 November 2015).

In Arsi, "Gallema Farmers' Cooperative Union became a 'market point' that needed 'improved seed'" (ibid.), with a malt barley seed shortage having been identified as a key 'bottleneck' to entry into the commercial/formal seed market arena. "We wanted to address why farmers are not growing malt barley and to organise and harmonise issues for the farmers" (ibid.). Gallema Farmers' Cooperative Union thus became a partner organisation to SHA in this malt barley initiative, which in turn led directly to the market agreement being signed between the farmers and Diageo later. This in turn allowed other actors to tie down that commitment in key locations, through establishing Memoranda of Understanding (MoUs) between the corporations, the ATA and barley farmers. This caused an immediate shift in seed practices, the formation of commercial clusters, the introduction of external seed and other essential inputs (pesticides and insecticides) and the opening up of market possibilities.

This decision has been important in determining seed futures in this locality according to SHA as "[i]n the past three year programme, this has led directly to increasing capacity from 100 kg bag of malt barley grain worth 600 Birr in 2012 to 100 kg bag of malt barley grain at 1200 Birr in 2015" (Interview with SHA Manager, 20 November 2015), with a significant shift in seed practice in this area, including not just malt barley grain production but premium prices for malt barley seed production also which I return to later. This has greatly increased the capacity of this cooperative and wider union structure as a market entry point geared towards formal seed production to improve the 'bottleneck of seed shortage' for the malt barley value chain.

A recent IFPRI (2015) report confirms that: "In the domestic market, the factory enjoyed monopsony power (one buyer but many sellers) over

---

[7] SHA Manager explains that two or more PCs make a union and two or more unions make a federation.

the malt barley sellers and, consequently, enjoyed some price setting power" (Rashid et al. 2015, p. 35). The report also points out that it had complete power of procurement for malt barley domestically and internationally in the preceding period. However, the entry of new market players—Heineken and Diageo—and a new malt factory, Gondar Malt, has significantly altered this structural arrangement and with increased competition in the sector the "Assela Malt factory has had to change its purchase price three times in 2014, with the initial price increased from 600–700 Birr per quintal to 900–1035 Birr per quintal" (ibid.). This was borne out by the interviews with the farmers in Digeluna Tijo worede (Interview with farmers at Senaboru PC, 24 November 2015).

One farmer explains that there was an informal PC before SHA came and that it was SHA that established more recognition for the cooperative and formalised it, which they assert has been to their benefit.

> The group is strength for us now. We experience sharing, solve problems together and can access different technologies and orient ourselves through the PC for the market. Now that we have information about market opportunities, we produce nowadays for the market, not really for food. Those of us who did not have a cow or an ox, now have improved livelihood because of membership of the co-operative. (Ato Tasfaye, Interview at Cooperative, 24 November 2015)

The farmers explain that they began with just 23 farmers in their cooperative in 2010, extending to 110 in 2015, with a projected rise to 140 farmers in 2016, and that being 'grouped' into a PC has provided them with new information and orientation for the 'market' and opened up opportunities for them to access technologies they otherwise would not get.

SHA confirmed that after selecting the Primary Cooperative (PC), here Senaboru PC, five PCs are chosen and a meeting is organised with the Executive Committee. Then attention turns to how individual farmers are selected for the malt barley value-chain programme. They are chosen in this case by external agents (SHA), with a strong government presence at the local level in the kebele and cooperative (Development Agents). This is a critical point in understanding how this 'transformation' of barley production has been so comprehensive in this case. The unique structure of governance and exercise of power within Ethiopia from the federal level down to the village is instructive in the context of how decisions are made regarding changes in seed practices in Oromia's barley belt.

The power structures run down from the federal level to the *killils*[8] or regions, which are governed by regional councils, whose members are directly elected to represent the districts or 'woredas'. The woreda is the third-level administrative division of governance in Ethiopia and this in turn breaks down to the local-level wards known as '*kebeles*' or neighbour-hood/peasant associations. The 'kebele' is the smallest unit of local government in Ethiopia, though it has "much less autonomy and decision-making power than the woreda, to which it is subordinate" (Rahmato 2008, p. 247). However, the kebele level does exercise administrative responsibility for core community functions, which are specifically relevant to this case as the kebele has a role in 'mobilising the people for community participation' (Bevan et al. quoted in Emmenegger et al. 2011, p. 10). The government-appointed agricultural development agent (DA) is a central member, amongst other key figures in each kebele's executive cabinet (Pankhurst 2008, p. 11).

More generally, the EPRDF operates a so-called one in five policy. This means that there is a party member for every five citizens, which acts as a 'silencing' system to its detractors.[9] Given an average of six per family, this means that the party has inserted itself more deeply into the community than any previous regime:

> the EPRDF has set 'a one to five' structure at the lowest level going down even to the family level ... thus able to reach every family in the country. This provides the ruling party with a huge potential for mobilising resources and the people for its developmental objectives ... but not without its drawbacks to the democratisation process. (Fiseha 2014, p. 72)

The Government DA at the Senaboru Cooperative in this case study plays an important role in disseminating the wishes of the Ethiopian Government. He explains:

> Each co-operative has to abide by the rules and regulations and guidelines set by the woreda though they have autonomy for setting their own internal

[8] Rahmato explains that "under the Federal system the country is divided into 9 major ethnic-based administrative units called killils in Amharic. The lowest unit is the kebele or sub-district and above it the woreda or district level" (Rahmato 2014, p. 223).

[9] According to Negussie (2014) the main opposition party *Adinet* "openly criticised the government at public rallies in recent times for using these forums to silence the public" (Negussie 2014, p. 185).

bye-laws. The Development Agent (DA) is a Government employee—the Government contribution to each cooperative. There are three DAs in each kebele. One for agronomy, one for animal husbandry and the third for natural resource management. The DA is also on the Executive Committee of the Cooperative. (Interview with DA at Gallema Farmers' Union, Arsi, Adam Sano, 24 November 2015)

This intensive web of power and governance which the EPRDF rigorously maintains allows federal policy to be realised rapidly at the local level. In this case, it meant that the development of a 'binding' contractual malt barley farming practice with Diageo was first agreed at the minutest level with government agents on the ground. This is confirmed by SHA when they explain the process of selecting the farmers who will plant new externally sourced improved seeds:

We focus on the farmers who do not have 'improved seeds', then the farmers who want or need, then build awareness and then they want to take the seed and produce grain. The Government Agent is there. The Executive Committee is there. Our focal person/employee is there from that area. They sit together and decide—to select the farmers who are under the PC. They see the list of the farmers and the Executive Committee know each of them. The Government DA knows them. After they select, they post the name of the farmers on the office of the Primary Cooperative: "For this year you are a member/ beneficiary of Self Help Africa. You get 'improved seed' from SHA and you multiply the grain". (Interview with Tadenya SHA Manager, 20 November 2015)

He goes on to explain that SHA also support Seed Producer farmers in this area cooperative providing them with basic seed:

They get 20% more than the market price, which is a premium because it is seed, not grain. They now produce seed from the Government Research Centre (EIAR), and they continue to produce this and sell it back to their co-operatives and the co-operative sells to other farmers. They now have standardised storage facilities. Now they have a seed cleaning machine, which is very important for selecting the purity of the seed. (ibid.)

This is a helpful clarification as it demonstrates that in two distinct ways farmers here are now introduced into the formal seed system (a) to produce malt barley grain from externally sourced malt barley seed, for the purpose of sale at a very good price under contractual agreement with a

TNC and (b) to produce malt barley seed (originating from corporate-sourced seed) for a premium price (20% extra) also for commercial purposes for selling back to the cooperative. This clearly signals a shift in farmer seed sovereignty. They have now given up half of their plots of land to externally sourced seed, thereby significantly reducing the land allocated to farmers' varieties and own food production, notwithstanding the benefits which may accrue from the changes, which the farmers themselves clearly saw as having value for them. Farmer sovereignty is thereby shared in new overlapping seed sovereignties, determined to a large extent by government/state agendas and their partnerships and agreements with other entities. The EPRDF-run State is directly involved in determining seed outcomes down to the lowest level of governance in the village/'kushet'.

Yet both SHA and the Government DA insist that the farmers maintain autonomy:

> The farmer still retains some independence as s/he can still make a contractual agreement with another private company, if the price is fixed and there is comparative advantage to do so … this negotiation can happen between the farmer directly with Diageo before the July planting and depending on the price offered, the farmer will decide whether to plant either food or malt barley, noting that generally the price of food barley is less than malt barley. (Interview with SHA Manager, 20 November 2015)

This was echoed by the Government DA, when he stated that the farmer still has a choice: "If he gets a better price, he can sell it in any market" (Interview with Government DA Adam Sano, 24 November 2015). Even the farmers asserted that this is the case, stating that they can access market prices now on their mobile phones and decide whether or not to plant more malt or more food barley depending on the respective price during the planting season. However, the idea they are exercising a free choice is debatable if the malt barley premium is double the price of food barley and if the contractual agreements that Diageo have agreed with the farmers are 'binding', and are sanctioned by the government, with government agents in every cooperative.

However, the farmers exercise agency in a different way. As we have seen, farmers maintain indigenous practices of crop rotation and traditional farming methods on at least half of their plot of land. They state "we will not stop growing our own food barley" (Interview with farmers PC

Senaboru, 24 November 2015). They speak in animated fashion about their own varieties and the myriad of uses and benefits accruing from them. Despite the area being targeted for commercial development of a malt barley value chain backed by powerful globalising forces, farmers retain a depth of knowledge and understanding of their need to practice and maintain access to a wide variety of locally adapted seeds for use throughout the year. This is essential for fulfilling year-round food needs which the uniformity required in monocropping production processes cannot cater for, especially on subsistence plots, but also for wider socio-cultural, local market and household needs.

The Government Development Agent and the SHA agronomist assigned to this cooperative did not regard the persistence of traditional seed practice for subsistence needs as unusual, despite new TNA-driven commercial agendas in this region. It was taken as a given. Three Government DAs are represented in every cooperative, as the Senaboru DA explained, and the EPRDF 'one in five' policy is effective down to the family level, so that in the Ethiopian context the indigenous seed practice would have to be both known and sanctioned by the government.

Dawit Alemu of the Ethiopian Institute of Agricultural Research (EIAR) provides an insight into the process. For Alemu:

> There is a definite push and pull of different actors in promoting and strengthening the formal seed system, but he emphasises that the informal seed system is still very good and very strong—still 100% for certain crops, such as barley which historically had very little contribution from the formal sector. (Alemu, Interview 2 December 2013)

This suggests that a twin-track seed policy is sanctioned by the government and given significant recognition by them in new seed legislation and on the ground here, unlike neighbouring Kenya, where parliamentarians noted the absence of any kind of recognition to traditional farmers and protection for local seed and food varieties. This is clear from the statement of the senior development manager at ATA when he stated: "we use our leverage to ensure that there is buy-in from the Government, as you cannot simply introduce something that is not palatable to the Government" (Interview with Yit Barek ATA, 20 November 2015). Specifically in relation to determinations relating to the malt barley decision he states:

Malt is less productive than food barley. Largely our farmers do not rely solely on one crop. They have a lot of diversity and do not give all of their land to malt ... the practice of farmer seed exchange without monetary value has been there for years. You have to respect farmers' rights too. It is also for food security. You cannot force everyone to buy seed. (Interview with Yit Barek, 20 November 2015)

In summing up, there is clearly a change in seed use and production arrangements. Externally sourced seed and contractual seed arrangements have been introduced with the signing of a MoU with Diageo. Land which once was totally allocated to non-commercially oriented traditional methods of seed use, production and distribution, which are central features of what we understand to constitute seed sovereignty, is now ceding some of that sovereignty to alternative seed pathways in new seed arrangements for malt barley production. Farmers expressed satisfaction with this new value chain based on better price and better quality it allows them. Though the degree of agency which they can assert in the decision-making process for making the switch is low, farmers in this locality still choose to use and produce own seed varieties of food and malt barley for subsistence production of grain and seed and the government know and accept this choice, even encourage it, despite the commercial value-chain model. In the next section I examine and interpret the motivations of the key actors involved in the decision-making process and how they determined the particular seed outcomes in relation to barley farming in this locality of the Oromia region at this time.

THE KEY ACTORS BEHIND THE CHOICE OF SEED PRACTICE

This section examines the role and motivations of the most influential domestic/state actors in this case. Dominated by the ruling EPRDF party, I examine how their political apparatus and determinations are affecting barley seed practices at the regional and local level. I also examine the role and motivation of the dominant external actors, IOs, TNAs and external state actors.

The Domestic Actors: The Role of the Ethiopian State

The previous section highlighted how the domestic structure of top-down governance in Ethiopia is central to the effective execution of seed practices at the local level. This is characteristic of what critics variously describe as a Maoist, Stalinist or Marxist-style (Clapham 2009) governance and is central to our understanding of the premeditated application of an unexpectedly differentiated seed programme on the ground in Oromia. I examine the behaviour of the different actors at the critical stages of the newly introduced barley value-chain decision-making process, looking first at the national level and then the local-level application. Two main questions are addressed, firstly, at a national level—why did the Ethiopian Government adopt a 'value-chain' approach when they were already embarked on their own developmental paradigm?; and secondly, having adopted this approach, why—at a local level—does the government allow the continuation and co-existence of the traditional, informal, non-commodified farmer-managed seed practice?

The National Level: The Ethiopian Style Developmental State and Emergence of the Value-Chain Approach

If Ethiopia's decision to embrace the Asian-type model of the Developmental State in 2006 marked a significant policy change in the country's agricultural industrialisation programme, the launch of the value-chain approach signalled that the Ethiopian State was now embedding itself in a technical and market-oriented paradigm with a very specific set of actors. Led by the World Bank, its chief funder, developing a malt barley value chain would be one of the key pilot projects of the newly formed, Gates-inspired ATA, testing the value-chain approach as a triggering mechanism for wider economic transformation.

The deepening of this more interventionist 'developmental state' model was based on a wider, transformative paradigm for the commercialisation of Ethiopian agriculture. It marks a significant policy shift from the earlier broad EPRDF-led ADLI strategy, which had dominated development discourse in Ethiopia in the 1990s and which focused primarily on 'bringing efficiency to small-holder agriculture', the main constituency base of the incumbent party. According to Ohno, neither ADLI nor the national Poverty Reduction Strategy Paper (PRSP) and PASDEP programmes of the early 2000s had succeeded in reflecting "a significant enough structural change such as crop diversification or productivity improvement" to trigger growth acceleration of the entire economy as some Asian countries had experienced (Ohno 2009, pp. 17–18) and lift people out of extreme poverty. By 2008/09, the Global South in particular was experiencing widespread unrest due to a global 'food crisis' and accompanying 'agflation' which saw 50% price hikes in some sub-Saharan Africa countries due to food commodity speculation given investment devaluation elsewhere (McMichael and Schneider 2011). The unrest this triggered would have been foremost in the minds of all actors as the new economic and development plans were being devised by the World Bank and the Ethiopian Government.

The World Bank, as a chief architect of global economic planning and a principal funder of Ethiopia's economic programme,[10] was insisting that a new economic, technical and developmental agenda was needed. Despite the Ethiopian State's "ambivalent attitude to economic liberalisation and the private sector" (Alemu 2011, p. 74), and tight state control over any transition, the World Bank held significant economic and political leverage over the government/state, which they now asserted. They were concerned that sub-Saharan Africa in general and Ethiopia in particular continued to experience a negative poverty loop, with agricultural imports increasing and their share of global agricultural exports decreasing,[11] (World Bank 2007, p. 2), despite success in overall positive growth rates in Ethiopia.[12] The World Bank saw the 'resurgence of interest in value chains' as a means to reinvigorate sub-Saharan Africa agriculture, stating that:

> experience shows that business concepts and methods are vital to developing value chains in an African context … where the specific focus is more on value creation, innovation, product development and marketing, thus incorporating supply, value addition, transactions and market linkages and sales. (World Bank 2007, p. 7)

2009 was a critical year for activating the value chain as a key policy mechanism and getting the EPRDF-run Government on board with what their leading economists referred to as a 'mindset change'. The pressure intensified on the Ethiopian Government at this time, as their main key donors and allies, who critically underpinned wider security (not just food) shifted their focus to the value-chain approach. The World Bank, which

---

[10] This included provision of its basic services and the Food Security net for seven million people.

[11] Agricultural exports from sub-Saharan Africa were predominantly primary commodities, with low value-added and decreased from 10% to 3% in the last four decades (World Bank 2007, p. 2).

[12] 2005–2015 Average growth rate of 7% per annum (PM Desalegn in ATA 2015, p. 3).

was now finalising its own new macro-economic plan for Ethiopia (AGP 2009–2011), and Ethiopia's most important geopolitical ally (notably the USA),[13] as well as new actors, BMGF with special interests in the seed space, all zoned in on 'value chains' as the key to unlock transformative change. They all identified malt barley as an agro-enterprise with the potential to deliver along the value chain—from production to delivery of a quality value-added beer product to the market. This would crucially link domestic and global markets, from smallholder farmers to the global beer market.

In the case of Ethiopia, the World Bank needed to persuade PM Meles Zenawi of this new approach before the new round of funding to the country was activated. The World Bank instigated many meetings between Meles Zenawi and leading economic thinkers such as Prof Joseph Stiglitz of Columbia University, Prof Dani Rodrik of Harvard University and Dr. Justin Lin chief economist of the World Bank at this time. Unsurprisingly, Zenawi was heavily influenced by their advice, though mindful that any change would have to be tightly managed by a state inherently distrustful of neo-liberal agendas, so as to avoid falling into a rent-seeking trap and anything that could stoke domestic or ethnic tensions that could undermine EPRDF dominance of state power. The economic advice was sobering for any sub-Saharan Africa leader at this time however. The central themes of the advice were clear—develop more trade openness, look to China and India as key export markets linking production and revenues with those countries' rising economic growth, decrease imports due to foreign exchange shortage and develop value-added agribusiness sectors linking domestic and global markets.

Ohno points out that in all likelihood the policy shift that Ethiopia agreed to at this time "was forced by the need to reduce imports under the severe foreign exchange shortage that Ethiopia has been facing since 2008" (Ohno 2009, p. 25–26). For this reason, the "emerging interest is in promoting import substitution industries" (ibid., p. 25). Rodrik specifically recommended a new industrialisation strategy arguing that "success depends on changing the mindset in which industrial policy is regarded to

---

[13] Christopher Clapham highlights how Ethiopia's swift backing of the US 'global war on terror' gave it "scope to promote its own agenda" (Clapham 2009, p. 181), effectively presented itself as "a force for stability in the region" (ibid., p. 190) and insulated it against possible loss of US support which it required for military protection of its border interests to the North in Eritrea and to the South in Somalia, all of which fed into US interests also to neutralise Islamist elements (ibid.).

a process of collaboration and problem solving with the private sector" arguing for "the need to move towards second-generation industrial policies that aim at both home market and exports" (Ohno 2009, pp. 25–26).

But even more significant for our understanding of the pressure that produced the outcome can be found in the footnotes of Ohno's paper. It refers to the fact that in May 2009

> The World Bank Executive Board approved the Protection of Basic Services (PBS) II, a budget-support type programme for Ethiopia … with the 'condition' [my emphasis] that the 'directional change' mentioned above be monitored for implementation. (Ohno 2009, p. 26, Footnote 7)

This was the most concrete lesson in realpolitik for Meles Zenawi. He was looking at his biggest donor—the World Bank—and its powerful financial institutional arm—the International Development Association (IDA), who were also embarking on a new round of funding for the wider economic programme, AGP1—flexing their financial and political muscle. The World Bank $540 million (loans and grants) PBS cheque allocated to health, education water and basic services 2009–2011 period was 'conditional' on a change in direction for the Ethiopian Government's 'mindset' that is to become more open for business. Twelve other donors would supplement the PBS II with an additional $737 million (Ohno 2009, p. 26). Ethiopia would change direction and open up to liberalisation, because the World Bank made it a condition of their funding of the fundamental service provision to the people, and the World Bank had the power to ensure that a dozen other key donors would follow suit. The Ethiopian Developmental State would work with business partnerships and share the space of new programmes which were already being devised—the value-chain commercial approach for key crop varieties, including barley was already underway.

Critically in the case of the proposition for a malt barley value chain, studies[14] were showing that demand was growing with annual growth rates of around 20%, "which is high given that it is almost double Ethiopia's real GDP growth rate over the same period" (Kaso and Guben 2015, p. 92). Domestic beer consumption was growing "by as much as 90% between 2002 and 2011" (FAO quoted in Rashid et al. 2015, p. 1), as the

---

[14] Some of these studies were the eight diagnostic studies funded and conducted by the BMGF (2008–2010) prior to the inception of the ATA in 2010.

population (now at 96 million) and income increased. Meanwhile, Ethiopia's import bill for malt barley had jumped from US $240,000 in 1997 to US $40 million in 2014, and was expected to rise to US $420 million by 2025. IFPRI point out "Given the country's balance of payment situation in recent years, this is an alarming trend" (Rashid et al. 2015, p. 1). Attracting the global corporate beer sector to stimulate a huge potential market, the kind of 'product distinction and branding' which Kaso and Guben identify as being increasingly important ingredients for market differentiation under globalisation (Kaso and Guben 2015, p. 84). Developing the entire malt barley value chain, substituting the import cost and producing a value-added beer product for domestic and export market was an attractive proposition and one which won favour with Zenawi and the EPRDF. Zenawi could accept it on the grounds of the DS model, which as Evans points out:

> Unlike orthodox socialism or communism that insists on state monopoly, developmental state accepts partnership with business elites (domestic or foreign) and other sectors of society. The ultimate goal of state development—is not the exclusive domain of the political elite or the bureaucracy—but should be a shared agenda that is based on a broad-based social and political coalition. (Evans cited in Fiseha 2014, p. 77)

In fact, sharing space with foreign business elites was preferable to domestic ones as it allowed the EPRDF to retain their hegemonic position within the state itself and use their effective one-to-five sub-kebele structural arrangement (Negussie 2014) to insert themselves at every stage of a tightly controlled value chain. This would be good for the optics of power, even if the reality was that powerful external actors/funders were in fact calling most of the tunes. Rahmato (2014) sees it as a policy bias in favour of foreign capital investment for large-scale commercially driven agriculture since the early 2000s (MOFED quoted in Rahmato 2014, p. 228), but which has invariably led to much publicised land grabs, particularly by Middle Eastern and Asian players and less participation by farmer-investors who would have much greater locally applicable knowledge (ibid., p. 240). Certainly the kind of 'economic determinism matched with the authoritarianism' of the DS model suited the tightly controlled 'value-chain' approach which gave greater latitude to EPRDF to maintain that system of cronyism and control. To that end, incentivising foreign direct investment and PPPs, whether through their "tax holidays, duty free imports for capital goods, grace periods of up to five years on land rents" (MoTI quoted in Alemu 2011, p. 74) were more attractive than sharing that 'liberalised' space with other domestic actors or the 'farmer/investor' base as Rahmato recommended (Rahmato 2014, p. 240).

It was into this space that corporate drinks giant Diageo inserted itself, when they announced their new investment at the World Economic Forum in 2012 stating:

> We will build a public-private partnership through which the Company will work with Ethiopian ATA to design and implement a barley contract farming strategy ... testing a pilot barley contract farming project with the aim to source 1000 metric tons (MT) barley from a substantial number of local small-holders in the first year. (Diageo Letter of Intent, signed at The Chicago Council on Global Affairs and World Economic Forum's Symposium on Global Agriculture and Food Security, 18 May 2012)

Diageo Ethiopia CEO confirms that Government of Ethiopia import-substitution policy was central to their agenda for entry to the Ethiopian market when they purchased Meta Abo Brewery in 2012 stating: "Meta is committed to work in line with the Government of Ethiopia's priority of substituting imports and saving foreign currency" (Phone Interview with Geoff Wallis, CEO Diageo Ethiopia 26 November 2015 and email questionnaire response 3 December 2015).

This was all in preparation for a much more important collaboration, devised at a global level and executed at a local level, now secured with the consent of the Government of Ethiopia. This resulted in the signing of the MoU with Diageo and ATA which was finalised in 2015 (Interview with Manager SHA, Tadenya, 20 November 2015). This particular PPP, and there are many others, and the ensuing MoU with farmers at the local level is set against the backdrop of the 2012 Global Agriculture and Food Security Symposium[15] and G8 Summit (Food Security Agenda).[16] It was

---

[15] New Investment was announced for Ethiopia at the Global Agriculture and Food Security Symposium in 2012. It was hosted by the Chicago Council on Global Affairs and World Economic Forum.

[16] The G8 NAFSN was launched in 2012 to mobilise private capital for investment in African agriculture. According to AFSA and GRAIN (2015, p. 4) "to be accepted into the programme, African governments are required to make important changes to their land and seed policies. The New Alliance prioritises granting national and TNCs new forms of access and control to the participating countries' resources, and gives them a seat at the same table as aid donors and recipient governments".

here that Paul Walsh, Global CEO at Diageo, signed a pledge committing to source up to 80% of its raw materials, that is malt barley, locally. This was also set against the domestic backdrop of an overall strategic plan to transform the barley sector not only to meet increasing domestic needs, but to become a regional exporter of malt barley also (ATA 2015, p. 68). From Diageo's perspective, this was an excellent corporate move into an expanding market with a growing population,[17] as the CEO of Diageo in Ethiopia attested. It appeared that they would be happy to comply with the idiosyncrasies of the Ethiopian approach.

It was also clear at this time that as well as the World Bank and US State interests in Ethiopian value-chain developments and the lucrative commercial rewards which liberalisation would bring, BMGF/AGRA were also developing closer ties with the Ethiopian State. The Ethiopian Government needed to keep on top of this new special relationship as much as possible. With an increasing number of powerful actors, the Ethiopian Government needed to ensure they were centrally involved in any unfolding value chain. This was especially relevant in the case of the mass of smallholder farmers who formed such an important base of their own political legitimacy and authority, but also held the potential for dissent, particularly in Oromia, where trouble was already brewing over land grabbing. Constructing a new elite agency with BMGF, namely ATA, and ensuring EPRDF control and dominance over it and the development of the value-chain agenda was essential.

What followed between 2009 and 2012 was consistent with the EPRDF regime's reputation for "brilliant" (Clapham 2009, p. 189) management of the international system, characterised by its adroit "ability" to present its own interests as "the answers to other actors' problems" (ibid.).

Firstly, in the Ethiopian case, the government needed to ensure that the execution of this new direction was both managed by them and, possibly more importantly from a public relations point of view, that it appeared to be controlled by them. This is important as Clapham (2009) points out that state power is increasingly dependent on maintaining an appearance

---

[17] Diageo has a connection with Africa dating back to 'the first recorded exports of Guinness to Sierra Leone in 1827 and employs over 5000 people on the African continent, one in four of its workforce worldwide' (Diageo press release 18 May 2012 [accessed online 7/10/2015]).

of popular support, particularly since the disputed 2005 elections (Clapham 2009, p. 185). In this context, Ohno's assertion that the Ethiopian Government insisted that the shift in emphasis was still "within their own fundamental (ADLI) policy" was instructive. Clearly, the World Bank saw it differently. It was calling it "a directional change in development policy" (Ohno 2009, p. 25), revealing a tension between the two actors. Secondly, and most importantly, as the execution of the policy found articulation in the value-chain strategy through the new ATA, we see the Ethiopian Government manoeuvring to ensure as much control as possible over the ATA itself. This was critical, given the increasingly close ties between World Bank and BMGF and AGRA interests, particularly in seed/crop value-chain developments.

To this end, the Ethiopian Government succeeded in ensuring dominance by their Executive and Council of Ministers on the Transformational Council of the ATA, despite only funding 5% of the new agency from domestic coffers. This meant that whatever programmes followed had to be sanctioned by the executive, even though certain powers were being shared with an emerging 'technocratic group', with new and powerful transnational actors exercising agency in the domestic seed space and pushing the 'new' economic/ industrial policy agenda through the value-chain strategy.

Of critical importance regarding the role of the newly established ATA was their acknowledgement at the outset that the agency would focus its direct support in implementing the solutions developed to key parts of the country identified by the Ministry of Agriculture within the World Bank's AGP, thus clearly indicating their choice of alignment.

Therefore, deferring to the AGP was not a simple random selection by the people leading and driving the new ATA, notably the Executive and the BMGF. It was again a strategic move determined by the importance of the actors involved. ATA's value-chain programme and ACC approach would dovetail with the already established work of their key donor and primary geopolitical ally through the World Bank/USAID globalising commercial agribusiness programme. This programme had already identified key, largely US agribusiness corporations as central players in rolling out 'transformational' agricultural plans in Ethiopia, where 'formalising' seed practices was central. USAID confirmed that they would parallel fund AGP, but work through the Government of Ethiopia in collaboration with MoA, ATA, Regional Government institutions as well as down to local/ kebele level (Interview with USAID 19 November 2015). Their main stated objective was to "shift farmers from their own open-pollinated varieties (OPVs) of seed to hybrid seeds for productivity purposes" (ibid.). USAID asserted that in the case of their programme with Du Pont/ Pioneer, for example, this was doing well, even if "it is taking a while for farmers to let go of their own varieties" (ibid.).

At a national level, the barley value chain and Agricultural Commercialisation Cluster (ACC) approach was entirely consistent with the objectives of the most powerful actors/allies, and now, crucially of the Ethiopian Government. However, as Clapham (2009) points out, the Ethiopian "regime was characteristically adept at pursuing its own interests and ideological commitment while presenting these in terms donors could accept" (Clapham 2009, p. 189). He identified their "key limitations on full acceptance of the 'Washington Consensus'", as firstly, the maintenance of state ownership of land, and secondly, the promotion, under the façade of liberalisation, of nominally private corporations that were owned by constituent parties of the governing EPRDF, especially by the TPLF[18] (ibid.). This all effectively ensured state control over the peasant population, yet even this is changing under World Bank AGP1&2.[19]

Kefale and Gebresenbet (2014) provide an important insight from their study on current large-scale sugar industrialisation in Ethiopia which is relevant here. They attest that the benefits of shifting to commercial agriculture "will accrue more at a national level, than at local levels. It has also political objectives. It provides the EPRDF the much needed legitimacy to prolong its stay in power" (Kefale and Gebresenbet 2014, p. 262).

It stands to reason therefore that if the overall picture of rising economic growth and continued investment of big companies like Diageo and Heineken and others continues, providing much-needed employment for the ever rising young, landless population, the EPRDF have the best possibility of staying in power.

To this end, Kefale and Gebresenbet (2014) state the EPRDF's "'developmentalism' is used as a political tool to increase economic development promising benefits to 'rural' communities, which the EPRDF hinges on to

---

[18] Tigrayan People's Liberation Front (TPLF) was the dominant element in the formation of the EPRDF.

[19] "Since 2003, the Federal Government has embarked on a new measure ... and rural land registration and certification is underway in four regional states of Amhara, Tigray, Oromia and SNNPR" (AGP 2015, p. 65), which amongst other things would "facilitate land use planning and management" (ibid.), all dismantling traditional community systems in favour of commercial ones.

get elected in consecutive elections" (ibid.). 'Rural' in this case, according to Kefale and Gebresenbet (2014), means those "small-holder farmers who live in the highlands of the four regional states (Amhara, Oromia, SNNPR and Tigray), which members of the EPRDF administer" (ibid., p. 263). This maintains a system of loyalty and vote management in the densely populated parts of the regions ruled by the coalition partners, from which all the gains of capital accumulation (which includes rents, taxes, profits, jobs and foreign currency) and reinvestment are redeployed to develop other mega projects. This, according to the authors, ultimately mostly benefits the highland region (ibid.). This system of patronage, led by the political elite is deeply embedded right down to the kebele level. It is increasingly open to corruption electorally[20] with implications for human rights.[21] Nevertheless, precisely because of the nature of this system, it is clear that all decisions, no matter how ambiguous or contradictory, are made to bolster the ruling party's hegemony in forging their political and developmental agenda and never occur without their knowledge and consent.

The Local Level: The Twin-Track Seed Practice

This section addresses the question why the Government of Ethiopia saw the benefit of the continuation of the traditional informal seed practice at the local level, despite adopting a value-chain policy. I highlight three interconnected themes: (1) the EPRDF's need to retain the support of the peasant class, which is inextricably linked to their own interest—regime survival; (2) the historical and continuing threat of food insecurity and famine; and (3) a heightened sensitivity to the need to protect the diverse genetic resource base especially in the context of climatic change.

1. Peasant Support Base and Regime Survival:

The Tigrayan-dominated EPRDF party "has its root in the mass lower class of rural farmers who led the armed struggle in the early 1990s and

---

[20] There has been much concern regarding the 2005 and 2010 elections. The EU claimed the 2010 election failed to meet international standards (EU quoted in Fiseha 2014, p. 84–85).

[21] According to Negussie, the main opposition party *Adinet* "openly criticised the government at public rallies in recent times for using these forums to silence the public" (Negussie 2014, p. 185).

who make up 85% of the population" (Fiseha 2014, p. 78). As a result, according to Teshome (2014), the party's policies and strategies have always been rural-focused—ADLI being the flagbearer for agriculture-development-led industrialisation since the 1990s when the EPRDF swept to power (Teshome 2014, p. 100). However, the adoption of the developmental state model by Meles Zenawi in 2006, with its inherent democratic deficit, alongside a dramatic change in direction by government in favour of GTP around the same time, meant that "family farming was no longer the darling of decision makers" (Rahmato 2014, p. 228). This would indicate that in fact the EPRDF Government wanted to move away from subsistence farming in favour of commercial agriculture and agro-processing. The controversial depeasantisation, decentralisation, villagisation and land grabbing, which some authors assert is led by the government (Rahmato et al. 2014), is part of this transition, opening up the country to investment and diversification of production. Profound change in Ethiopia's seed system is now underway. But the transition is tightly controlled by the government, intent on maintaining EPRDF-led State dominance over other seed actors in the process, especially their traditional farmer support base.

To this end, peasant subsistence agriculture and the public seed programme continue under state-directed control of the entire seed system. Alemu (2011) highlights this continuation through his examination of two contrasting seed policy initiatives currently being pursued, which explains the evidence of a twin-track seed policy in the Central Highlands. He highlights the federally driven 'Crash Seed Multiplication Programme', which emphasises improved seed productivity, while simultaneously promoting decentralised, locally-run 'farmer-based seed production and marketing (FBSPM) schemes', mainly targeting open-pollinated crop varieties (OPVs) (Alemu and Tripp quoted in Alemu 2011, p. 72). This FBSPM plays a vital role in the national seed system, according to Alemu, as it is the main source of raw seed for the formal public seed enterprises, which all the regional seed enterprises in turn rely on (ibid.). Significantly:

> these schemes improve the possibility of seed production of locally demanded varieties and crops for which there is no commercial interest. There is also an increased possibility of producing and marketing seed within communities, so reducing costs. (Yonas Sahlu et al. quoted in Alemu 2011, p. 72)

This is critical to a cash poor country like Ethiopia.

Deeply connected to this, it must be remembered that the Government of Ethiopia "command considerable power over the small-holder population, through a land system which is based on state or public ownership, and private property in land is not allowed by law" (Rahmato et al. 2013, p. 127). Thus, this land policy, which is based on 'Usufruct Rights',[22] enshrines access to land in the constitution, but does not address three key aspects according to Teshome (2014), namely (1) tenure security; (2) farm size and fragmentation; and (3) the issue of land markets (ibid., p. 101). This impoverishment certainly restricts farmers' possibilities and their room to manoeuvre in determining their own futures, and is widely accepted in the literature that it impacts on the ability of the people to resist top-down decisions on agronomic or other practices/policies coming from federal level. It is also central to the criticism of many authors regarding the state's primary role in land grabbing, and accompanying concerns over displacement. This calls into question the various assertions by MoA and ATA personnel that their motivations are entirely guided by their wish to protect farmers' rights. MoA state: "The Government wants to protect the farmers. That is a priority agenda of the Government" (Interview with Mekonnen, 11 November 2015), and similarly ATA say:

> You have to respect farmers' rights too. It is also for food security. You cannot force everyone to buy seed. (Interview with Yit Barek Semeane, Director Seeds System, Ethiopian ATA, 20 November 2015)

However, it is more likely that it is in the government's interest to maintain food barley and traditional methods because it cannot afford monetarily to do otherwise at this juncture and it needs state-led hegemonic practices to prevail. This signals government weakness rather than strength.

2. Threat of Food Insecurity and Famine:

"The politics of hunger looms large" in economic policy and in political decision-making, according to Alemu (2011). This is not least because the last two regime changes came on the back of horrific famines

---

[22] "Land users have use rights over the plots they hold, and these plots cannot be sold or mortgaged" (Rahmato et al. 2013, p. 127).

and considerable loss of life. The continuation of farmer seed practice[23] was also considered essential in the context of the threat of food insecurity and famine. The ADLI strategy of 'betting on the small-holder' was coming in for some criticism. Leading Ethiopian social scientist, Dessalegn Rahmato (2008), argues that subsistence smallholders lacked the capacity to realise the earlier ADLI vision, when 37% of them cultivate less than or equal to 0.5 hectares (Rahmato, quoted in Teshome 2014, p. 101, fn 5), which Rahmato calls "starvation plots" (ibid.), while 87% subsist on less than 2 hectares (ibid.), a point echoed in interviews with World Bank specialists (Interview with Woet Soeur, Senior Social Protection Specialist, World Bank, 30 November 2013).

Alemu (2011) gives an insight into why this may be so in referencing the 'peculiar' application of a uniquely Ethiopian version of the Green Revolution there. Apart from the ambivalence towards the private sector, which has been synonymous with EPRDF rule, Alemu highlights the reticence within government of total liberalisation as there is a genuine fear that "if we just liberalise today without any capacity, the whole market will fail" (Interview with Alemu, 2 December 2013). This he cites as the reason for "holding on to the vibrant informal seed system which is also very good" and operates external to liberalised commercial markets. For Alemu:

> Despite the push for hybrid seeds and commercial systems, because the vast majority of the seed system is reliant on the informal seed system (97%), and because access is a major factor inhibiting most very poor small-holder farmers engaging in the formal sector, as well as the challenge of climate change, they (the government) have to be very careful regarding narrowing choice. If there is only one choice, there is no choice. (Alemu D, Interview 2 December 2013)

The fear of a narrowing of seed choice to the vast population of subsistence farmers is deep rooted with the recent memory of famine and food

---

[23] Ninety-seven per cent of the population still rely on heterogeneous, non-commodified, informal seed systems. This is "'local seed', carried over from the previous harvest, either by the farmers themselves (through the traditional on-farm selection process, whereby the farmer identifies next years seed stock while it is still maturing in the field and gives it special protection) or by buying from preferred seed stock kept by other farmers in the same locality" (Alemu 2011, p. 70, quoting FAO and WFP 2008). Even the formal sector remains predominantly in the public sector, unlike other countries, where 'formal' usually implies private-sector ownership or control.

insecurity, and ensuing political turmoil, following the horrific loss of life during the Ethiopian famine of the 1970s and 1980s, which in the case of the 1980s eventually numbered close to one million dead.

This emphasis was evident when the new Seed Proclamation was being introduced. Hassena et al. lamented the "overdue emphasis on food crops, while the issue of export crops (horticulture in particular) were not considered" (Hassena et al. 2016, p. 87). Their study identified the source of this tendency directly in the office of the PM and his experts, indicating that despite the intense pressure from a myriad of other very powerful actors, the EPRDF-run Government were insistent on maintaining an emphasis on food crops produced by the majority subsistence farming population for now.

The informal subsistence farmers, who number 97% of the population, place a great value on maintaining and enhancing seed diversity on their own plots for such food crops. They value this practice as it enables them to stagger food supply, planting some varieties which can be harvested early and some late varieties for harvesting later in the season, thus ensuring food security for their household throughout the year. Local practices are synonymous with heterogeneous polycultures, the opposite of homogenous monoculture cropping for market systems. Poor subsistence farmers in harsh climatic conditions particularly require this kind of heterogeneity, as they have multiple uses for different varieties, required at different times. Barley alone can be used for animal feed, for diverse culinary uses, for thatching and for different spiritual and religious occasions, which are deeply ingrained in the socio-cultural lifeworld of the Ethiopian people. There is an immense knowledge and value attached to these local practices, many of which are specific to the different regions and different agroecologies. Farmers do see the value, monetary and otherwise, of trying new ways, new seeds and new market opportunities as outlined above, but the reality of the climatic and economic conditions they work within ensures that the non-commoditised seed sharing and saving practices must continue to ensure food security well into the future.

3. Diversity

A third reason for ensuring a differentiated application of seed practice at the local level lies in a strong commitment by the Ethiopian State to maintaining its seed sovereignty. They therefore support the continued practice of maintaining heterogeneity in seed/crop varieties in the country's vast agroecologies, which became a feature of seed programmes since the worst famines of the 1970s and 1980s. This led to the establishment in 1976 of the EBI gene bank. It marked an important moment of reap-praisal of Ethiopia's unique status as a Vavilovian centre of diversity and a strong attachment to its role as a globally important repository of plant genetic resources (PGRs). EBI, as a publicly accountable institution of state has been influential within government circles, across the African continent and beyond as referenced in the last chapter. For its founder, Dr. Melaku Worede, the logic of the unique Ethiopian programme is borne out of that intense learning period, has been based on:

> Raising productivity without compromising the broad adaptive gene com-plex which is inherent in the plasticity of the landrace varieties, which our farmers, observe, select and adapt in-situ, and now, critically, conserve through utilisation. (Interviews with Worede 23 November 2013 and 17 November 2015)

Most importantly, he states that this is supported by the government and valued by them, not just as central to the sovereign heritage of the Ethiopian people, but as a buffer against calamitous shocks which the nature of climatic challenges present, especially to smallholder farmers. The Ethiopian State has actively engaged in this germplasm exchange with the farmers since, distributing seed to key stakeholders (farmers, breeders, researchers and research centres) for wider dissemination and use through-out Ethiopia's vast agroecologies throughout the agricultural year. Regarding the twin-track approach, Dr. Regassa Feyissa points out that:

> The present model is not only limiting our capacity of thinking, but more importantly cannot work in our fragile conditions. Climate change is a witness for us that we do not have any guarantee to discard anything that may be useful to us—which is why our small-holder farmers need to ensure the continued evolution in nature of their seeds, not because they are poor, or because they want to do conservation, but because out of farm, means disappearance of the resource base. (Interview with Feyissa, 26 November 2013)

The government maintains an attachment to the values of seed sover-eignty, at least for now. They provide continued support for the protection of the indigenous 'resource base' through the twin-track approach and other initiatives, notwithstanding the fact that they removed the EBI man-date to regulate access to genetic resources (Seed Proclamation 381/2004 Art 6, as quoted in Feyissa 2006).

CONCLUSIONS FROM THE OROMIA STUDY

The Oromia barley case study reveals the World Bank as a key driver in determining macro-economic changes to the Ethiopian Government's order of business. It highlights the extent of the reach of World Bank influence on pushing key economic and political agendas, right down to the 'irreducible core of agriculture'—the seed—and directed deep changes in agricultural orientation including seed policy and practice during this time with the introduction of externally sourced commercial malt barley seed in new contractual arrangements with subsistence farmers for the first time. This in turn facilitated many other actors, internal and external, to come in behind their agenda. It dictated the end of the domestic ADLI programme and ensured the shifting of GTP1&2 to embrace the World Bank/AGP and a pro-commercial/corporate value-chain approach in a selected sub-sector. It effectively created this shift in 'mindset' by making it a conditional part of its financial arrangement for the provision of basic public services to the Ethiopian Government, which is necessary to sustain the population. This was set in the context of the World Bank's already established alignment with USAID through its seed and marketing programme AGP/AMDe. This allowed the ATA to facilitate and develop the accelerated commercial cluster programme to introduce new commercial seed production arrangements between the corporations like Diageo and the Ethiopian Government for key value chains. This is precisely what occurred in the case of the switch from farmers' seeds for predominantly food barley to corporate seed-derived malt barley production on a considerable portion of farmers' land in a locality in Oromia.

Seed, once geographically grounded and shared in distinct agroecological zones, is now also externally sourced, corporate-owned, a mobile artefact, commodified and contractually shared with Ethiopian farmers who are bound in a value chain. This is changing and diversifying their livelihood, bringing some benefits according to the farmers. However, though this exposed a certain weakness and dependency on behalf of the EPRDF, the discovery of a twin-track seed practice in the locality suggested some unexpected and contradictory arrangements. The application of the value chain in this area is by no means totally globalised, despite Diageo's 'letter of intent' for this pilot barley contract farming project being first signed at the Chicago Council on Global Affairs and World Economic Forum Symposium in 2012. Traditional seed practices continue alongside it—a key example of 'glocalisation'. This insertion of the global inserted into the local is a key indicator of the kind of hybrid globalisation transformationalist scholarship insists is a key feature of this approach. The 'fragmegrative' state is sharing the seed space in this case study with significant others. These important interests/actors are determining the fate of seed sovereignty now, which, though still largely held in state and farmer hands, is by no means fixed or congruent, just as transformationalist scholarship would suggest.

Traditional seed practices continue to co-exist and are accommodated, albeit by an authoritarian state, attempting to control its mass peasant power base. Despite the decision of corporate drinks giant Diageo to invest in the barley value-chain/import-substitution programme designed by the ATA and implemented by SHA on the ground in Oromia, it is clear that the Government of Ethiopia maintain tight control over the publicly-dominated formal and informal seed system, which for now are not of interest to external actors. It is administered by them at a federal level and executed with tight state observance at a local level. There is also a clear rationale that the country cannot totally risk its seed sovereignty to either the vagaries of the weather, or the vicissitudes of a neo-liberal market, or a potentially volatile peasant class, with a deep socio-cultural attachment and knowledge of their diverse agroecosystems. This adds to the contingent nature of this era of globalisation, mirrored in continued government commitment to maintaining community agrobiodiversity. Therefore, the government, for now, encourage farmer seed practices to continue on a certain percentage of their state-owned land, even in the targeted zones for globalised seed practices in the Central Highland region of Oromia.

At the outset of this case study, one could have been forgiven for thinking that hyperglobalism was the best fit for interpreting the process afoot, but it could not explain the wider empirical findings. Equally, one could not say, as sceptics would, that the state was a dominant actor, despite its obvious adroit manoeuvring to maintain patrimonial arrangements and their own political hegemony, which the value-chain approach entailed, as well as their obvious knowledge and approval of local actors' alternative seed practice. In this scenario, the specific distinctions which the transformationalist approach brings best explain the process. As this approach suggests, globalisation is affecting the state not eroding it. The state is showing a capacity to reassert itself in the new shared space with globalising actors, making strategic choices, assisted by its authoritarian approach. Yet, there is a disaggregation, as the domestic seed policy space is being diffused and core functions are increasingly being shared with other actors such as the World Bank, USAID and new transnational actors such as BMGF and in this case, Diageo.

The application of the value chain in this area is by no means totally globalised, nor is the state necessarily the main player, despite its central involvement. Other important actors are at the table and seed sovereignty is certainly shifting to these new actors, but in an ambiguous way, and for myriad reasons, with positive and negative flows. Seed sovereignty is still largely held in state and farmer hands, but is by no means fixed or congruent, just as transformationalist scholarship would suggest.

REFERENCES

Abay, F., Bjornstad, A. and Smale, M. 2009. Measuring on farm diversity and determinants of barley diversity in Tigray, northern Ethiopia. *Monoma Ethiopian Journal of Science*, 1(2), pp. 44–66.

Abay, F., de Boef, W. and Bjørnstad, Å. 2011. Network analysis of barley seed flows in Tigray, Ethiopia: supporting the design of strategies that contribute to on-farm management of plant genetic resources. *Plant Genetic Resources*, 9(4), pp. 495–505.

AFSA and GRAIN. 2015. *Land and seed laws under attack, who is pushing changes in Africa?* [Online]. Available from: www.grain.org [Accessed 22nd January 2016].

AGP. 2015. *Agricultural Growth Program (AGP-II): Social assessment report final.* Addis Ababa, Ethiopia: World Bank.

Alemayehu, F. 1995. *Genetic variation between and within Ethiopian barley landraces with emphasis on durable resistance.* PhD thesis. Wageningen University, Landbouw.

Alemu, D. 2011. The political economy of Ethiopian cereal seed systems: state control, market liberalisation and decentralisation. *IDS Bulletin*, 42(4), pp. 69–77.

ATA. 2015. *Agricultural transformation agenda progress report covering 2011–2015 in the GTP I period.* Addis Ababa, Ethiopia: ATA.

Ayenew, M. 2014. The growth and transformation plan: opportunities, challenges and lessons. *IN:* Rahmato, D., Ayenew, M., Kefale, A. and Habermann, B. (eds.) *Reflections on development in Ethiopia: new trends, sustainability and challenges.* Addis Ababa, Ethiopia: Forum for Social Studies, pp. 3–30.

Clapham, C. 2009. Post-war Ethiopia: the trajectories of crisis. *Review of African Political Economy*, 36(120), pp. 181–192.

Emmenegger, R., Keno, S. and Hagmann, T. 2011. Decentralization to the household: expansion and limits of state power in rural Oromiya. *Journal of Eastern African Studies*, 5(4), pp. 733–754.

Ethiopian Census. 2007. *Census-2007 Report.* Addis Ababa: Central Statistical Agency. Available at: http://www.csa.gov.et/census-report/complete-report/census-2007. Accessed 14 March 2019.

Feyissa, R. 2006. *Farmers' rights in Ethiopia: a case study.* Lysaker, Norway: The Fridtjof Nansen Institute.

Fiseha, A. 2014. Development with or without freedom? *IN:* Rahmato, D., Ayenew, M., Kefale, A. and Habermann, B. (eds.) *Reflections on development in Ethiopia: new trends, sustainability and challenges.* Addis Ababa, Ethiopia: Forum for Social Studies, pp. 3–30.

Hassena, M., Hospes, O. and De Jonge, B. 2016. Reconstructing policy decision-making in the Ethiopian seed sector: actors and arenas influencing policymaking process. *Public Policy and Administration Research,* 6(2), pp. 84–95.

IFPRI. 2015. *The barley value chain in Ethiopia.* Washington DC: International Food Policy Research Institute.

Kaso, T. and Guben, G. 2015. Review of barley value chain management in Ethiopia. *Journal of Biology, Agriculture and Healthcare,* 5(10), pp. 84–97.

Kefale, A., and Gebresenbet, F. 2014. 'The Expansion of the Sugar Industry in the Southern Pastoral Lowlands', in Rahmato, D., Ayenew, M., Kefale, A. and Habermann, B. (Eds). *Reflections on development in Ethiopia. New trends, sustainability and challenges.* Addis Ababa, Ethiopia: Forum for Social Studies, pp. 247–268.

McMichael, P. and Schneider, M. 2011. Food security politics and the millennium development goals. *Third World Quarterly,* 32(1), pp. 119–139.

Negussie, S. 2014. *Reflections on development in Ethiopia. New trends, sustainability and challenges.* Addis Ababa, Ethiopia: Forum for Social Studies.

Oakland Institute. 2017. *Down on the seed: the World Bank enables corporate takeover of seeds.* California, USA: Oakland Institute.

Ohno, K. 2009. *Ethiopia: Political regime and development policies* [Online]. Available from: www.grips.ac.jp/forum/af-growth/support_ethiopia/document/Jun09_DD&ADLI_10E.pdf. [Accessed 12th February 2013]. GRIPS Development Forum: Tokyo.

Pankhurst, A. 2008. *Enhancing understanding of local accountability mechanisms in Ethiopia: protecting basic services project.* Addis Ababa, Ethiopia: PBS II Preparation Studies, Revised Summary Report.

Rahmato, D. 2008. *The peasant and the state: studies in agrarian change in Ethiopia, 1950s–2000s.* Addis Ababa: Custom Book Publishing.

Rahmato, D. 2013. Food security and safety nets: Assessments and challenges. *IN* Rahmato, D., Pankhurst, A., and van Uffelen, J-G., *Food Security, Safety Nets and Social Protection in Ethiopia,* Addis Ababa, Ethiopia: Forum for Social Studies, pp. 113–146.

Rahmato, D. 2014. Large-scale land investments revisited. *IN* Rahmato, D., Ayenew, M., Kefale, A. and Habermann, B., *Reflections on development in Ethiopia. New trends, sustainability and challenges.* Addis Ababa, Ethiopia: Forum for Social Studies, pp. 219–245.

Rahmato, D., Pankhurst, A., and van Uffelen, J-G. 2013. *Food security, safety nets and social protection in Ethiopia*. Addis Ababa, Ethiopia: Forum for Social Studies.

Rahmato, D., Ayenew, M., Kefale, A. and Habermann, B. 2014. *Reflections on development in Ethiopia. New trends, sustainability and challenges*. Addis Ababa, Ethiopia: Forum for Social Studies.

Rashid, S., Abate, G.T., Lemma, S., Warner, J., Kasa, L. and Minot, N. 2015. *The barley value chain in Ethiopia*. International Food Policy Research Institute (IFPRI): Washington, DC.

Spielman, D.J., Byerlee, D., Alemu, D. and Kelemework, D. 2010. Policies to promote cereal intensification in Ethiopia: the search for appropriate public and private roles. *Food Policy*, 35, pp. 185–194.

Teshome A. Development programs and the post-MDG agenda. *IN* Rahmato, D., Ayenew, M., Kefale, A. and Habermann, B. 2014. *Reflections on development in Ethiopia. New trends, sustainability and challenges*. Addis Ababa, Ethiopia: Forum for Social Studies, pp. 97–129.

Worede, M. 2011. Establishing a community seed supply system: community seed bank complexes in Africa. *IN:* Li Ching, L., Edwards, S., Scialabba, N. E. (eds.) *Climate change and food systems resilience in sub-Saharan Africa*. Rome. Italy: FAO, pp. 361–377.

World Bank. 2007. *Using value chain approaches in agribusiness and agriculture in sub-Saharan Africa: a methodological guide*. Washington, DC: World Bank.

World Bank. 2013. *Agribusiness indicators: Kenya*. Washington, DC: Agriculture and Environment Services, The World Bank.

CHAPTER 7

Reshaping Seed Sovereignty

There are four new important insights arising from this book. Firstly, this study applies different approaches to globalisation to the issue of seed sovereignty for the first time. Secondly, this research applies these theoretical approaches to an empirical study in sub-Saharan Africa. Thirdly, this book provides the first in-depth study of new Kenyan and Ethiopian seed laws for the first time and also provides the first local study of seed sovereignty in Ethiopia. Fourthly, it proposes that transformationalist theory is a useful framework for assessing the changing nature of seed sovereignty in sub-Saharan Africa and that it has the potential to be applied more generally.

This chapter brings together the main elements of the study. It begins by providing a synopsis of the key elements of the book. It then draws together the overall conclusions, indicating where these findings fit into the literature and the significance of the contribution the work brings to this area of enquiry, offering some conclusions and recommendations arising from it.

REVISITING SEED SOVEREIGNTY AND GLOBALISATION

Chapter one introduced the debates surrounding globalisation's challenge to traditional, state-centred, notions of sovereignty. Central to this debate is the increasing role of other players in global political arenas, namely

© The Author(s) 2019
C. O'Grady Walshe, *Globalisation and Seed Sovereignty in Sub-Saharan Africa*, International Political Economy Series,
https://doi.org/10.1007/978-3-030-12870-8_7

Table 7.1   Competing approaches to globalisation revisited

	State	*IOs*	*TNAs*
Hyperglobalists	Deterritorialised	IOs set worldwide rules	They operate worldwide
Sceptics	State-centred	IOs operate by state-to-state bargaining	Vehicles for state interests?
Transformationalists	Some states are stronger than others—US hegemony, Chinese in Africa	In IOs some states are more important than others and also some states are more important in some areas	Some are more dominant than others and some are important in some areas, but not all

IOs, TNAs, TNCs and NGOs. There is now a myriad of new global actors. This has given rise to a rich new literature theorising globalisation. I adapted Held and McGrew's (2007) work and identified three competing approaches: hyperglobalism, scepticism and transformationalism. These varied in relation to their interpretation of three key areas, the role of the domestic state, IOs and TNAs. For each approach, I identified the different stances taken on each of these three issues (see Table 7.1).

Hyperglobalism sees the state eroded of its original core functions. Whether globophobe or globophile, it sees the state system as a redundant force, hollowed out by new actors who exercise power. Sceptics disagree, arguing that states still matter. Strong states still exercise power and old geopolitical alignments still determine outcomes. Transformationalists insist that both approaches are too simplistic. They see the state as a disaggregated player, sharing the space with a multiplicity of new actors, but retaining its principal juridical attributes, albeit in conflictual and ambiguous ways.

Chapter 2 traced the evolution of thinking on food security to the emergence of food sovereignty into the lexicon. The study then focused on seed sovereignty. It identified how the three approaches could be applied to seed sovereignty. This is the first time that this kind of exercise has been undertaken (see Table 7.2).

The aim was then to see which, if any, of these approaches captured the empirical reality of seed sovereignty. I chose to study the most recent seed laws in two sub-Saharan Africa countries, the Kenya Seed and Plant Varieties Amendment Act (SPVAA 2012) and Ethiopia Seed Proclamation (782/2013). This was the first time that such a comparison has been

Table 7.2   Three perspectives on globalisation and seed sovereignty

	State	*IOs*	*TNAs*
Hyperglobalists	Deterritorialisation—Transboundary nature of seed mobility Erosion of state/public role in seed policy/programme	UPOV, WTO, TRIPS, AoA set rules backed by World Bank, IMF. Harmonisation of laws	Core duties and functions derogated to others—especially TNCs Global penetration mergers and acquisitions Corporate control—Monsanto, Du Pont, Gates, AGRA and others
Sceptics	State is still the main driver of policy choices and policy implementation	IOs do what big states tell them	TNAs still look to strong 'Northern' states who determine outcomes
Transformationalists	'Multiple equilibria' State is a disaggregated player but still has a role in certain aspects. Glocalisation—local initiatives	Ambiguities within World Bank and FAO re IAASTD report Contradictions within UN role and substance CBD, IPCC	Multidimensional not unidirectional Technological and other innovation can go either way Homogeneity and heterogeneity

conducted and provided the first in-depth study of each law. I also conducted a within-country study of a new barley value-chain programme in Oromia, Ethiopia. This was the first local study of seed sovereignty in Ethiopia.

The three approaches generate expectations about the motivations and actions of the key actors in the seed space and their role and motivation in determining seed policy sovereignty in distinct locales. I break them up into four headings, namely (1) role of external actors—states/IOs; (2) role of TNAs; (3) role of TNCs; and (4) specific adjustments to laws and policy, as detailed in Table 7.3.

The Kenyan case revealed a clear example of hyperglobalism. The state, through the legal process of drafting, clearly facilitated its own withdrawal from core, previously public, seed functions in the face of global commercial interests. The powerful seed actors who intensified their business in an increasingly liberalised Kenya at this time were the net beneficiaries as seed

Table 7.3 Expectations of the three approaches in relation to the three case studies

	Hyperglobalists—expected evidence
Role of external states and key IOs	1. Strong evidence of high-level meetings between key IO reps and presidents/prime minister and key ministries prior to publication of seed bills and launch of value chain, especially regarding UPOV and TRIPS, and global standards for seed
Role of key TNAs	2. Strong evidence of meetings between philanthropic foundations, notably AGRA/BMGF, post-2006 prior to publication with senior state representatives and officials of key national agricultural institutes, and absence of CSO and farmer groups from the process
Role of key TNCs	3. Strong evidence of TNC dominance over state actors, causing a dislocation of key public institutions in the final seed law or practice
Key adjustments in text of bills/laws	4. Strong evidence that the wording of the law or construction of a new seed practice will include direct references to issues that IOs, philanthrocapitalists and seed TNCs have been calling for, that is the bill itself and the value chain approach

	Sceptics—expected findings
Role of external states and IOs	1. Strong evidence of external state involvement (mainly US as the global seed hegemon) behind other actors/actions, signifying geopolitics behind seed laws and new seed practices
Role of domestic state actors	2. Strong evidence of state officials leading new seed initiatives and centrally involved in all stages of the drafting and implementation of each seed bill through to publication
Role of TNAs/ TNCs	3. Strong evidence that TNAs/TNCs do what big states tell them
Adjustments in key texts of bills/laws	4. Critical variations in bill/law signifying state differentiation despite globalising forces

	Transformationalists—expected findings
Role of external states and key IOs	1. Strong evidence of meetings between the state and global actors such as the World Bank and big state actors (mainly US as the global seed hegemon), and certain disaggregation of the state's role in seed arena which would be reflected in contradictory outcomes in final legal text and seed practice
Role of key TNAs	2. Strong evidence of new TNAs such as BMGF/AGRA and TNCs in new, complex multidimensional arrangements, not unidirectional and to be reflected in the final seed law or practice
Adjustments in seed bills/laws	3. Strong evidence of state retreat in some areas ceding seed sovereignty to globalising forces/standards, but retention of seed practices in other areas in ambiguous and complex interactions between the global and local forces

Table 7.4   Kenyan Seed Law SPVAA 2012 adheres to hyperglobalist expectations

	Evidence
Role of external states	1. Strong evidence that Kenya was chosen as pilot country by USAID-funded ASARECA, ECAPAPA and COMESA to lead seed harmonisation in favour of the strictest globalised seed rules UPOV with new regime for PBRs and PVP and deliberate exclusion of informal seed networks in the region
Role of IOs	2. Strong evidence of close connection between President Kibaki and World Bank, who were major donor and supporter of his Vision 2030 favouring seed and agricultural changes. Three government agriculture ministers supported application of stricter UPOV rules throughout the law-making process
Role of TNAs/TNCs	3. Strong evidence that TNAs dominated by key seed TNCs pushed for seed regulatory changes through establishment of 'shadow task forces' including STAK and PBAK throughout the process. All of their proposals were enacted in SPVAA 2012. CSOs and farmer bodies were not consulted in the drafting process
Adjustment in bill/law	4. Strong evidence that the process led directly to the dislocation of the formerly public seed regulator Kephis, opening it to privatisation—a key feature sought by commercially driven globalising forces

sovereignty shifted to them. When tracked against the different approaches, this becomes very clear, as the Kenyan case conforms to the hyperglobalist model on every count as represented in Table 7.4.

The research into the making of Ethiopian Seed Proclamation 782/2013 revealed a more nuanced and highly ambiguous process and, in turn, a differentiated seed outcome. The empirical findings suggest that the process did not conform to either hyperglobalised or sceptical interpretations of globalisation, when tested against the different approaches, though indicating signs of both (see Table 7.5).

The case study of the newly established malt barley value chain in the Oromia in Ethiopia localised study revealed a most contradictory and a highly differentiated seed outcome on the ground. Despite clear globalising forces, notably Diageo, the World Bank, BMGF/ATA and World Economic Forum, this study showed a clearly transformationalist outcome, with the deliberate retention of a twin-track seed practice. Corporate and indigenous seed practices co-exist within the locality (see Table 7.6).

Table 7.5   Ethiopian seed law 782/2013: A transformationalist approach

	Evidence
Role of external state/IOs	1. Strong evidence that US state and related IOs are key geopolitical ally to Ethiopian state. US seed/chemicals TNC, Du Pont/Pioneer run the USAID externally derived corporate seed programme in Ethiopia, which is directly linked to the World Bank AGP seed agenda in Ethiopia. Yet they were not directly involved in the drafting process
Role of key TNAs	2. Strong evidence of highly ambiguous relationship between PM/executive of EPRDF and private philanthrocapitalist organisation, BMGF, who funded and carried out eight key diagnostic studies, which led to the setting up of an extra-parliamentary ATA, which at a critical juncture took over the entire drafting of the new seed law, dislocating other key actors in the process, notably EBI—a designated accountable institution of state with specialist role in conservation, utilisation and access and benefit sharing of Ethiopia's vast seed repository
Role of domestic state actors	3. Strong evidence that the EPRDF-led government, as the main domestic state actor, retained many of its principal juridical attributes amongst diverse actors/interests. Despite being an obviously disaggregated player in the process of making a new seed law, and the exclusion of CSOs and farming bodies, the law ambiguously reflects them more than in the Kenyan case
Adjustments in the seed law	4. The sharing of the disaggregated seed space led directly to a highly differentiated three-tiered seed law, giving recognition to the formal, informal and intermediate standards for seed quality, and special exemption for 97% smallholder farmers from having to conform to the new law

KEY FINDINGS FROM THE CASE STUDIES

The empirical application of the different accounts of globalisation to changing regulatory frameworks and seed practices in sub-Saharan Africa confirmed that seed sovereignty is indeed affected by global forces, but in different ways in different localities. The research explored the nature of this polycentric new architecture in the seed space. One case, Kenya, is an example of a hyperglobalised seed law. It grants proprietorial rights over improved uniform seeds to TNC/commercial interests and dislocates the domestic state regulatory authority, Kephis, opening up a previously public function to privatisation, a critical feature of hyperglobalisation. This is unsurprising given the carte blanche afforded by the Kenyan government to relentless pressure from key donors/interests (WB, US government

Table 7.6 The Oromia case: Malt barley value chain—A transformationalist approach

	Evidence
Role of external states and key IOs	1. Strong evidence of (IO) World Bank as the key player pushing value-chain approach at global level, yet state retained a role in sovereign seed practices, albeit as a weakened and disaggregated player. The outcome is reflected in the findings as corporate malt and indigenous food barley growing in 'segregated plots' in the same locality
Role of key TNAs/TNCs	2. Strong evidence of 'glocalisation'. Diageo letter of intent for PPP first signed at global forum in Chicago. Signed MoU with ATA, SHA, Farmer's Union and local farmers. Brought their own seed and introduced contractual farming at local level, yet farmers remain as seed producers for own seed varieties on their own plots as local, informal seed practices prevail
Key adjustments in actual seed practice	3. Local actors designated by key agents of an international aid agency, SHA and Government Development Agents at co-op level. Farmers expressed positive opinion about the switch to externally sourced malt barley variety (mainly due to good price arrangements and guaranteed markets) for the grain and in some cases seed produced from the new method. They also express the value for their own varieties which they maintain on farm

bodies, AGRA) and seed TNC interests and by its exclusion of key civil society groups and farmer bodies from meaningful engagement with the drafting process. However, extending the comparative study to neighbouring Ethiopia revealed a markedly less globalised contemporaneous law, which gives recognition to differentiated seed practices and includes an exemption favouring farmers' seeds and the informal seed system. The domestic state still mattered, but not in a way that sceptics would suggest. The process of drafting the new Ethiopian Seed Proclamation revealed complex manoeuvring between a myriad of diverse actors, donors and domestic state actors in ambiguous ways, yet also excluding CSOs and farming bodies. Despite the obvious dominance of certain actors (notably the office of the PM at state level) throughout the drafting process, what emerged was a situation where a conflictual political space was shared and seed sovereignty was ceded in some clauses to other actors/funders/interests, but not in all. Similarly, the within-country case study in Oromia provided another example of this kind of ambiguity. Here we find a commercially oriented globally inspired malt barley value-chain initiative, announced in Chicago at the World Economic Forum, but which reveals a twin-track seed programme in the same location. Corporate, externally sourced Diageo seeds and new contractual malt barley farming is underway alongside the segregated plots of indigenous food barley and deeply ingrained informal farmer/community seed networks. The Ethiopian case studies revealed the state as a disaggregated player, not hollowed out as hyperglobalists would expect, nor the central fulcrum of power as sceptics would suggest. Paradoxically, despite the state showing strength in retaining considerable sovereign seed control, state weakness is exposed in the face of multiple pressure sources, including external pressures, internal pressures and a volatile climate, prone to famine.

These case studies confirm that on balance, it is more likely that state sovereignty is neither being eroded as hyperglobalists suggest, nor reasserting its power and dominance as the primary actor as proponents of the sceptical school of thought claim. Instead, as the transformationalist perspective asserts, national sovereignty is increasingly divided and shared between local, national, regional and global authorities. What emerges are 'overlapping sovereignties' in complex new arrangements and heightened conflict and insecurity at all levels accompanying these new 'transgovernmental relations'. Globalising forces are transforming the state's seed policy role, opening up multiple and ambiguous pathways. The effect is potentially highly conflictual, compounding problems as the shift from government to polycentric governance in the seed area is profound. It is, as anticipated, even more conflictual given the skewed regulatory effects and asymmetrical distributional consequences of this new wave of globalisation, particularly in an African context. The exclusion of CSOs and representative smallholder farming organisations from any role in the decision-making process is a significant feature of all three case studies. The top-down manner of decision-making is another key feature, which indicates where power does and does not lie in this new polycentric space. The formation of shadow task forces and fast-track legislation in the Kenyan case and the construction of an extra-parliamentary Agricultural Transformation Council, so closely tied to an unaccountable TNA, BMGF, in the Ethiopian case, are signals of what the new disaggregated space of seed sovereignty looks like. The inclusion of these 'new' transnational philanthrocapitalist actors is significant. Due to their wealth and power, they now enjoy unprecedented access to domestic governments, which facilitates a strategic position for them in new contradictory arrangements of seed governance, where these new actors have unprecedented leverage in determining new seed laws and policies in different jurisdictions. They stand to benefit enormously from these ventures. Unlike public companies that are answerable to shareholders, or IOs, even the World Bank, these private philanthrocapitalist foundations are answerable and accountable to no one. Nevertheless, the case studies highlight that these uncertain new arrangements hold both positive and negative actual and potential outcomes, exactly as transformationalists assert. The Ethiopian cases are a case in point, where the state is adapting to sharing the space of power and therefore to sharing sovereignty with other key transnational actors, leading to both integration and fragmentation occurring simultaneously and with complex and contradictory results. Despite not including farmers in the drafting process, the Ethiopian seed law included a critical exemption for smallholder farmers and a differentiated seed system was accommodated, despite the intensity of globalising forces involved directly in the drafting of the law. The existence of a twin-track seed practice in the Oromia case reveals a similar contradictory accommodation. This gives practical credibility to the transformationalist's assertion that this era is more inclined towards 'organised heterogeneity than strict homogeneity', that globalisation has no central premise. The 'fragmegrative state' is the distinct sign of the new polymorphous arrangements representing the simultaneous integrative and fragmenting nature of the new dense web of seed governance. These new transworld actors are now playing a key role where the state once occupied sole jurisdiction. Testing the different accounts of globalisation against these empirical studies allowed this picture to be revealed more clearly, deepening our insight into how seed sovereignty is being affected now in two important East African countries.

APPROACHES TO SEED SOVEREIGNTY

Adapting and applying three competing approaches to understanding the process of globalisation on sovereignty provided new evidence on three main counts.

Firstly, it suggests that this model of analysis constitutes a useful heuristic device that can be applied to empirical studies of seed sovereignty. This way of working, though not absolute, makes the identification of actors, interpretation and analysis of key events, actor motivation and subsequent outcomes more accessible, thus greatly assisting the comparative method in such studies.

Secondly, the research provided evidence of the kind of change transformationalist authors assert is the hallmark of globalisation in this era. This finding is all the more contentious in the context of sub-Saharan Africa and the charges of a vastly unequal globalisation occurring there, accompanied by an inevitable political turmoil. It also usefully exposed the polymorphous nature of new arrangements of transnational governance across the seed space at this time in two key African countries. The domestic state is undoubtedly a disaggregated player in restructured, ambiguous, multidimensional arrangements of power and influence. This is likely to compound problems of conflict and governance in an already highly contested space, particularly in the already vulnerable communities of sub-Saharan Africa, which are in the frontline of worsening climate change.

Thirdly, this point specifically suggests that at the sub-Saharan Africa level, transformationalism is the most appropriate characterisation because we see variation across the cases. This approach provides the most useful framework for assessing the myriad aspects which changes in seed sovereignty in sub-Saharan Africa brings, and suggests that there are grounds for considering that this framework can be applied more generally.

These findings are significant. Academics, particularly the broad spectrum of scholars involved in the international colloquia at Yale (2013), the Hague (2014) and Vittoria Gasteiz (2017), notably, Scoones and Thompson (2011), Edelman (2014), Kloppenburg (2014), Patel (2009), Murphy (2014) and Carolan (2012), as well as practitioners on the ground, including Abay et al. (2011), Di Falco and Chavas (2009), Feyissa (2006), Worede (2011), Alemu (2011), Odame and Muange (2011), Louwaars and De Boef (2012), have been calling for new models of analysis to assist us in understanding changing seed policies and to sharpen our focus on an unfolding food and seed politics planet-wide. The academy has asked (Edelman 2014; Murphy 2014)—who is the sovereign in food sovereignty? This study has problematised this vexed question through an empirically grounded study, which reveals how different actors are affecting the current exercise of seed sovereignty. It finds that sovereignty is increasingly shared and in some cases changing hands in the seed context. It exposes the significant actors who have been omitted and excluded from the decision-making process, not least the majority farming populations, who remain the primary seed and food producers across the continent and who stand to lose most when and if policies fail. The UN bodies have been increasingly concerned at the neglect of farmers' needs, with the Special Rapporteur on Food noting particularly the loss of traditional methods of seed saving and exchange and the dangers of the loss of biodiversity to "uniformisation encouraged by the spread of commercial varieties" (De Schutter, 2009). The UN Human Rights Council has presented a draft Declaration on 'small farmer rights to seeds' (Art 19) (passed in November 2018) and a separate article on the right to biodiversity (Art 20) (Coordination Sud 2017), specifically to address this issue. Similarly, the IAASTD report of 2009 highlighted the risk posed by such uniformisation. Crucially, enhancing agrobiodiversity through conservation and equity in access and benefit sharing is central to the ongoing work of the CBD and its important supplementary agreement, the Nagoya Protocol which has entered into force in October 2014, "giving greater legal certainty and transparency for both providers and users of genetic resources" (Nagoya Protocol 2010).[1] All of these initiatives make this comparative case study of seed laws and practices in different jurisdictions highly relevant and useful to these important international objectives. It also crucially offers a robust model of analysis that can be replicated in the study of other jurisdictions to broaden our view of what is happening elsewhere, particularly in other agrarian societies in the Global South. The process of drafting and the final outcome of recent seed laws in sub-Saharan Africa and the application of a globalised malt barley value-chain reveal important complexities in power arrangements which we urgently need to understand in this new disaggregated seed policy space. Precisely because of the contradictory nature of the findings and the differentiated outcomes in different locations, this study provides what transformationalists Held and McGrew (2007) argue is a possibility of reshaping or even reforming globalisation.

Paradoxically, despite the accusation of enclosure-capitalism which value-chain arrangements can represent (McMichael 2012) and the authoritarian nature of the EPRDF-led 'developmental state', Ethiopia provides an ambiguous, but nevertheless more flexible and locally appropriate interpretation favouring a differentiated seed system in both of the Ethiopian case studies. This adds to the growing literature, which has been calling for such a varied application in domestic settings, under the 'sui generis' clause of UPOV and TRIPS for example (Munyi et al. 2016; Munyi 2015; De Jonge 2014; Louwaars and De Boef 2012; Alemu 2011; Dutfield 2011; Scoones and Thompson 2011; and Tansey 2011). These diverse local articulations are important in light of the unnecessarily oppressive interpretation of UPOV being adopted under duress in many

[1] https://www.cbd.int/abs/about/#objective

(particularly poorer) jurisdictions now. Alemu (personal interview 2013) stated that one choice is no choice. The IDS Bulletin special edition on seed politics in 2011 similarly stated that the 'one-size-fits-all' seed solution and increasingly donor-led agenda setting was highly problematic and that there was a need to highlight opportunities for reshaping the terms of the debate (Scoones and Thompson 2011). This study provides an important contribution to that debate. Leading Ethiopian scientists, like Melaku Worede, Regassa Feyissa and others, who have considerable experience and integrity in addressing these complex issues, must be included at the seed policy-making table at every stage. This is essential; particularly when new seed laws and seed practices are being assessed and developed. For Worede, "narrowing the choice is untenable in countries that require a broad adaptive resource base, and which is inherent in the plasticity of landrace farmers' varieties" (Interviews with Worede 23 November 2013 and 17 November 2015), which are synonymous with seed sovereignty and which a majority of rural dwellers still rely on, especially in climate-vulnerable regions. This is backed up by other recent scientific studies in the region (Abay et al. 2011; Di Falco and Chavas 2009; Worede 2011; Feyissa 2006; De Boef et al. 2012) and by a plethora of global institutions such as CBD, IAASTD, International Treaty on Plant Genetic Resources for Food and Agriculture (especially relating to ABS and farmers' rights), Office of the UN Special Rapporteur on Food, IPC of UNFAO, Nagoya Protocol, Cartagena Protocol, none of whom were consulted and whose views are therefore not reflected in the new seed laws in two of the most climate-vulnerable countries on the planet.

This study provides both evidence and a model of analysis that could inform future work in this area. The transformationalist approach provides a practicable framework to address the contradictory elements of the inter-penetrative processes which occur when global forces interact with, or even 'collide' with local realities in these situations. In this study we see where polycentric seed governance is rapidly intensifying, with mixed results. It is giving rise to ambiguities and contradictions within global institutions themselves, as evidenced in the case of the World Bank/WHO-commissioned IAASTD report (IAASTD 2009), and the increasing pressure coming from within UN FAO to recognise farmers' rights, agroecology and biodiversity conservation. However, these contradictions, in keeping with transformationalism, offer an important avenue of hope for future work in reshaping critical aspects of the present globalising seed narrative. This research adds to the rich emerging literature and sign-posts the need for better regulatory mechanisms which ensure broader and transparent participation in the seed law and policy-making processes. This will be important as we can expect more contradictory and conflictual features to emerge within the seed space with the rise of Brazil, Russia, India, China and South Africa (BRICS), particularly Chinese interests in seeds. State-owned ChemChina's takeover of Syngenta in 2017 is worthy of watching as China's influence and investment in Africa in particular, dwarfs most other players. Applying this model of analysis will be useful in assessing how seed sovereignty fares when different external state actors are involved. Already there are signs of critical differentiation, which are worthy of study. Chinese and Russian seed sovereignty movements are also showing very different characteristics than the more vocal Latin American-styled peasant movements such as LVC. Since food security and seed security are issues intrinsic to every society on the planet and will remain centrally important, the application of this approach is a useful tool to deepen our understanding of the fate of seed sovereignty in distinct locations offering important insights for policy-makers and practitioners alike. This study indicates that domestic seed sovereignty could be unnec-essarily compromised, resulting in decisions that gravely undermine the rights of local populations and the diverse ecosystems they and we rely upon for survival on a finite planet. Extending this study and model of analysis to other jurisdictions now, particularly other climate-vulnerable locations would broaden our understanding of what is occurring and assist in developing more equitable outcomes. To this end, recent seed laws, especially in sub-Saharan Africa, including in Uganda and Tanzania, could be and are likely to be revisited and repealed. This is already underway in Kenya. This study will assist this process within these countries in pushing for inclusion and legal recognition for differentiated seed systems, to rein-state the importance of agroecology and to ensure that farmers' rights are enshrined in law. This model of analysis can assist in vetting all new seed laws and seed policies presently under construction, informing parliamen-tarians and practitioners of appropriate participatory practice which includes all voices, particularly those which promote farmers' rights as encapsulated in the UN Declaration of 19 November 2018[2] regarding the

---

[2] http://www.ip-watch.org/2018/11/23/un-committee-adopts-landmark-declaration-reinforcing-peasants-rights-seeds/

rights of peasants and people working in rural areas, following from the draft UN Declaration on Peasant Rights and Biodiversity Rights (2012) Articles 19 and 20 pertaining to peasant/farmers rights to seeds and to biodiversity. This study provides not just an important intellectual contribution to this unfolding debate but demonstrates important practical implications and applications, offering us a good place to begin to reshape and rethink globalisation and its relationship to seed sovereignty. This research suggests that transformationalism is a useful framework for assessing changes in seed practices in sub-Saharan Africa and that it has the potential now to be applied more generally.

A Call for Decisive Action

The conclusions drawn from this book give serious pause for thought theoretically. However, the implications go well beyond the academy. The presentation of differentiated seed policies, law and practice across the case studies along with the conclusion that transformationalism best captures the reality on the ground signals an opportunity for action. This can directly inform activists and practitioners on the one hand and national governments and IOs on the other to act more decisively in determining more equitable outcomes in the seed space.

Decisive action in this regard is sorely needed. This is not least because of the increasingly volatile and potentially devastating effects of climate change on seed and food production, particularly in climate-vulnerable countries such as those in sub-Saharan Africa. Transdisciplinary calls for biodiversity protection and farmers' rights and farmer knowledge to be included in policy and decision-making can find a most immediate and practical focus in all new seed legislation coming before national parliaments and in policy-making processes at the international and national levels. The reform of those seed laws is essential now to bring countries, particularly the most vulnerable ones, in line with best environmental and climate-proof practice.

The historic UN Declaration of 19 November 2018[3] regarding the rights of peasants and people working in rural areas marks the culmination of years of work. It follows landmark work such as the UN Millennium Development Goals of 2000, the UN Millennium Ecosystem Assessment

---

[3] http://www.ip-watch.org/2018/11/23/un-committee-adopts-landmark-declaration-reinforcing-peasants-rights-seeds/

Reports of 2005 and 2010, the work of the Sustainable Development Goals since 2015 and the 2009 ground-breaking report of the IAASTD (2009). This latter report stated categorically that 'business as usual is not an option', arguing for genetic variability in the plant varieties being grown as the best buffer against climate change. These reports had all been set in the backdrop of other important protocols and conventions such as Access and Benefit Sharing of the Convention on Biological Diversity (ABS of CBD) of 1992 and the Nagoya Protocol of 2010, which was enacted in October 2014. New domestic seed laws and policies are a litmus test in critically important climate-vulnerable countries of how these important declarations and transdisciplinary calls are being implemented and adhered to or not. They reveal the international and domestic influences and influencers determining seed practice, and therefore food and agricultural futures. They also alert us to the vested interests involved in the seed space and their modus operandi, particularly in directing globalising seed law to their own benefit.

Given the findings, I identify four priority areas for immediate action:

1. Revision of international seed law and policy—especially UPOV to include farmers' seeds and rights to save, use and exchange own seed in farmer breeding programmes. UPOV needs to be updated in line with the fundamental Principles of Nagoya 2010 and ABS of CBD (1992).

International laws such as UPOV 91, WTO regulations, TRIPS and other intellectual property laws need to reflect the other important international laws and policies such as Nagoya and ABS of CBD (1992) and the recent UN Declaration on the Rights of Peasants and Biodiversity of 2018. In effect, this is to ensure that environment and development concerns are joined up to the highest level of our understanding of the needs, particularly those of marginalised communities within an international legal framework. Given that UPOV 91 is aimed predominantly at advanced economies, with highly uniform commercial agricultural production systems, it has long been argued that the LDCs dominated by majority smallholder subsistence farmers such as Ethiopia should not have to adhere to such standards. This issue must be urgently addressed. These strict compliance rules are becoming bargaining chips for WTO inclusion and are a competitive advantage in business. This undermines seed sovereignty, but also puts food security at risk. International law must ensure that this is not allowed to occur. Leading reports commissioned by the World Bank and WHO, such as IAASTD and UN bodies have cautioned against this strict compliance. Leading experts on seed and food security have also called for it. Access to diverse seed preferred by farmers is essential. Leading climate scientists are calling for it, too. All agree—keep diversity alive. In-situ evolution of locally adapted crop varieties is a way forward. Indigenous knowledge is valuable. Farmers' rights and farmers' seeds must be acknowledged, listened to and acted upon within emerging legal frameworks. Globalising seed laws must reflect the full range of law in this arena in this era. They must not be based on market concerns or driven by commercial interests. To this end the role of IDLO must be scrutinised. The role of the BMGF on this body is problematic in the context of the evident commercial gain for the philanthrocapitalist body in the context of making new seed laws and policies in countries like Kenya and Ethiopia where they are centrally involved.

2. Evaluation of all existing and proposed domestic seed laws and policies to reflect the basic principles of seed sovereignty, particularly in climate-vulnerable zones with large subsistence farming communities to ensure agrobiodiverse seed and food security.

Domestic seed law needs to reflect emerging best practice in the seed space, which includes protection of farmers' rights and biodiversity rights as per the Nagoya Protocol, IAASTD and the UN Declaration on the Rights of Peasants and Biodiversity of November 2018, amongst others. These reports and international agreements do not reflect rights for their own sake, but are set in the context of essential food security and livelihood needs. The conclusions here identify that the domestic state can afford to assert its seed sovereignty at the domestic level to a much greater extent. It must not be held hostage to corporate/commercial interests. It can be bolder in ensuring that the plant genetic resources of the state are neither obliterated nor appropriated by external forces.

Keeping options open is essential in the context of climate change, particularly for domestic states in sub-Saharan Africa. This will entail ensuring that seed sovereignty is reflected in new seed laws, including revising those elements of laws, such as in Kenya, which have outlawed seed saving and exchange amongst peasant farming communities.

3. Expansion of community seed banks planet-wide as part of pro-grammes for the evaluation and enhancement of farmers' varieties to stimulate the use of germplasm resources that are already locally adapted. This could become central to national programmes for future-proofing for climate change.

The articulation of seed sovereignty at the local or community level is one of the most exciting and potentially transformational responses emerging worldwide at present. It is finding traction on every continent. These range from the kitchen gardens of quiet food sovereignty in Russia, where dacha gardens produce 40% of Russia's food, to the regeneration of land-race varieties of wheat across Italy led by farmers, bakers and scientists collaborating with government, to widespread community seed-banking programmes in Ethiopia. Increasing examples of seed sovereignty are emerging from China, Asia, Latin America, with slow food movements across Europe, community and NGO-led seed banks, and increased national interest in heritage seed conservation and use. New community-led governance structures are providing important evidence-based local results, of how people are fundamentally able to rethink and redesign their seed/food systems around agroecological principles, with positive results for yields as well as conservation and wider social capital gains. These counter-narratives to a one-directional globalisation, which locks farmers, communities and countries into industrial food and farming systems, are essential for the transition that is needed. Maintaining seeds in the hands of local actors has wider value. In this sense glocalisation (i.e. the insertion of the global in the local and vice versa), a central feature of transforma-tionalist thinking, offers the opportunity for a profound reform of globalisation in the seed space in this era. As above, this will also entail drafting new seed policies, which reflect and enshrine seed sovereignty going forward. It can build on innovative work already underway in domestic programmes of community seed banking, which can be administered regionally and locally. Such agroecological projects are becoming hugely important and increasingly funded though nowhere near enough to make them viable and meaningful agricultural enterprises. A recent survey found 84 dedicated community seed-banking programmes are already underway in Europe alone (DIVERSIFOOD 2018). They point the way to a diverse seed future and must be enhanced, expanded and supported fully now nationally and internationally.

4. Development of sustainable evolutionary breeding systems for the generation of genetic variability through 'in-situ' (on-site) conservation programmes of farmers' varieties on smallholder farms linked to agricultural research organisations across countries and regions and agroecologies.

This development would serve to ensure the continuation and protection of local and national stewardship as opposed to privatised corporate incursions into the public seed space and the take over and ownership of the planet's plant genetic resources. This development would be extremely important in the context of climate change in general and sub-Saharan Africa in particular. Ethiopian scientist Melaku Worede clarifies that it is this in-situ conservation of farmers' varieties on smallholder farms which is "providing a valuable option for conserving crop diversity especially in many parts of Africa subject to drought and other stresses because it is under such environmental extremes that variations useful for stress-resistance breeding are generated. In the case of diseases or pests, this allows for continuing host-parasite co-evolution" (FAO 2011, p. 364). This change would ensure wider food security, not only through the inherent evolutionary dynamic of adapting materials in extreme conditions, but by ensuring the broadening of the range of seed material available—what Worede calls the 'inherent plasticity' of this type of system. It would also, most importantly, provide planting material to ensure food security that could also be fed back into research and breeding programmes at the institutional level for crop improvement. This has long been regarded as essential in maintaining global food security and there are countless examples of its reach in times of crisis. Maintaining such dynamic systems would be a universal contribution of immeasurable value.

CONCLUSION

The problems addressed in this book regarding the fate of seed sovereignty are clear. Sovereignty over seed and food futures is under threat at multiple levels. Some are biophysical, others institutional and policy directed. The interaction between global institutional and transnational actors, on the one hand, and domestic and local actors, on the other, is central to understanding the politics of seed sovereignty in the context of international relations. It is politics which determines who gets what, where, when and why. It thereby determines who the winners and losers are in that political process. The forensic examination of the making and enactment of new seed laws illuminates both the actors involved and the nature of their interaction and reveals the process that determines outcomes in the seed space. The winners in this case determine who will potentially own and control the biological systems and agriculture of the future. The stakes are very high indeed, not least for those who lose out. These stakes extend way beyond seed sovereignty and offer important signposts for the future from both a policy and practical point of view.

REFERENCES

Abay, F., de Boef, W. and Bjørnstad, Å. 2011. Network analysis of barley seed flows in Tigray, Ethiopia: supporting the design of strategies that contribute to on-farm management of plant genetic resources. *Plant Genetic Resources*, 9(4), pp. 495–505.

Alemu, D. 2011. The political economy of Ethiopian cereal seed systems: state control, market liberalisation and decentralisation. *IDS Bulletin*, 42(4), pp. 69–77.

Carolan, M. 2012. *The sociology of food and agriculture.* Earthscan Routledge: New York.

Coordination Sud. 2017. *The right to seeds: a fundamental right for small farmers!* Paris, France: Coordination Sud.

De Boef, W.S., Thijssen, M.H., Shrestha, P., Subedi, A., Feyissa, R., Gezu, G., Canci, A., Ferreira, M.A.J.D.F., Dias, T., Swain, S. and Sthapit, B.R. 2012. Moving beyond the dilemma: practices that contribute to the on-farm management of agrobiodiversity. *Journal of Sustainable Agriculture*, 36, pp. 788–809.

De Jonge, B. 2014. Plant variety protection in sub-saharan Africa: balancing commercial and smallholder farmers' interests. *Journal of Politics and Law*, 7 (3), pp. 100–111.

De Schutter, O. 2009. *Seed policies and the right to food: enhancing agrobiodiversity and encouraging innovation report of the Special Rapporteur.* A/64/170 United Nations General Assembly [Online]. Available from: http://farmersrights. org/pdf/righttofood-n0942473.pdf [Accessed 23 February 2013].

Di Falco, S. and Chavas, J. 2009. On crop biodiversity, risk exposure, and food security in the highlands of Ethiopia. *American Journal of Agricultural Economics*, 91(3), pp. 599–611.

DIVERSIFOOD. 2018. DIVERSIFOOD Report: Community Seed Banks in Europe. DIVERSIFOOD stakeholder workshop in Rome on 21 September 2017. Available at: http://www.diversifood.eu/community-seed-banks-in-europe/. Accessed 26 December 2018.

Dutfield, G. 2011. Food, biological diversity and intellectual property: the role of the International Union for the Protection of New Varieties of Plants (UPOV). *Intellectual Property Issue*, Paper No.9. Quaker United Nations Office.

Edelman, M. 2014. The next stage of the food sovereignty debate. *Dialogues in Human Geography*, 4(2), pp. 182–184.

FAO. 2011. *Potential effects of climate change on crop pollination*. Rome: FAO.

Feyissa, R. 2006. *Farmers' rights in Ethiopia: a case study*. Lysaker, Norway: The Fridtjof Nansen Institute.

Held, D. and McGrew, A. 2007. *Globalisation/anti-globalisation: beyond the great divide*. 2nd edition, Cambridge: Polity Press.

IAASTD. 2009. *Agriculture at a crossroads: a synthesis of the global and sub-global IAASTD reports*. Washington, DC: Island Press.

Kloppenburg, J., 2014. Re-purposing the master's tools: the open source seed initiative and the struggle for seed sovereignty. *Journal of Peasant Studies*, 41(6), pp. 1225–1246.

Louwaars, N.P. and De Boef W.S. 2012. Integrated seed sector development in Africa: a conceptual framework for creating coherence between practices, programmes, and policies. *Journal of Crop Improvement*. 26(1), pp. 39–59.

McMichael, P. 2012. The land grab and corporate food regime restructuring. *The Journal of Peasant Studies*, 39(3–4): 681–701.

Munyi, P. 2015. Plant variety protection regime in relation to relevant international obligations: implications for smallholder farmers in Kenya. *The Journal of Intellectual Property*, 18(1–2), pp. 65–85.

Munyi, P., De Jonge, B. and Visser, B. 2016. Opportunities and threats to harmonisation of plant breeders' rights in Africa: ARIPO and SADC. *African Journal of International and Comparative Law*, 24(1), pp. 86–104.

Murphy, S. 2014. Expanding the possibilities for a future free of hunger. *Dialogues in Human Geography* [Online], 4(2), pp. 225–228. Available from: http://journals.sagepub.com/doi/pdf/10.1177/2043820614537166 [Accessed 26th September 2014].

Nagoya Protocol. 2010. *Nagoya Protocol on Access to Genetic Resources and the Fair and Equitable Sharing of Benefits Arising from their Utilization to the Convention on Biological Diversity*. Quebec: Secetariat of the Convention on Biological Diversity, United Nations Environmental Programme.

Odame, H. and Muange, E. 2011. Can agro-dealers deliver the green revolution in Kenya? *IDS Bulletin*, 42(4), pp. 78–89.

Patel, R. 2009. What does food sovereignty look like? *The Journal of Peasant Studies*, 36(3), pp. 663–706.

Scoones, I. and Thompson, J. 2011. The politics of seed in Africa's green revolution: Alternative narratives and competing pathways. *IDS Bulletin*, 42(4), pp. 1–23.

SPVAA. 2012. *Seeds and Plant Varieties (Amendment) Act, 2012 (No. 53 of 2012)*. Nairobi: Kenya Gazette Supplement, No. 217.

Tansey, G. 2011. Whose power to control? Some reflections on seed systems and food security in a changing world. *IDS Bulletin*, 42(4), pp. 111–120.

UN International Declaration of Peasants' Rights. 2012. *Final study on the advancement of the rights of peasants and other people working in rural areas.* Geneva: United Nation A/HRC/AC/8/6. Available at: http://www.wphna.org/htdocs/downloadsfeb2013/2012%20Declaration%20of%20Peasants%27%20Rights.pdf. Accessed 14 March 2019.

Worede, M. 2011. Establishing a community seed supply system: community seed bank complexes in Africa. *IN:* Li Ching, L., Edwards, S., Scialabba, N. E. (eds.) *Climate change and food systems resilience in sub-Saharan Africas.* Rome. Italy: FAO, pp. 361–377.

Appendix

List of Interviewees

Interviews in Kenya

Dr. Robin Buruchara, Africa Regional Director of International Centre for Tropical Agriculture (CIAT), which is part of the Consultative Group for International Agricultural Research (CGIAR) consortium, and key institutional partner of Forum for Agricultural Research in Africa (FARA).

Senator Hassan Omar Hassan, Secretary General of the Wiper Democratic Movement of Kenya and member of Coalition for Reforms and Democracy since 2013 with Orange Democratic Movement (ODM). He is a lawyer and former Commissioner and Vice-Chairperson of the Kenyan National Human Rights Commission (KNHRC).

Dr. Bonny Khalwale, MP for Ikolomani (2002–2013) and Senator for Kakamega (2013–2017).

Hon Dr. Wilber Ottichilo Khasilwa, MP and member of the Orange Democratic Party for the Emuhaya Constituency.

Dr. Sally Kosgei (by phone), former Minister for Higher Education and Science and Technology (2008–2010) and Minister for Agriculture (2010–2013).

Daniel Maanzo, MP for Makueni. A lawyer and member of the 29-member agricultural committee, which drafts legislation.

© The Author(s) 2019   259
C. O'Grady Walshe, *Globalisation and Seed Sovereignty in Sub-Saharan Africa*, International Political Economy Series,
https://doi.org/10.1007/978-3-030-12870-8

Daniel Maingi, Director of the Kenyan Food Rights Alliance (KEFRA) and Growth Partners Africa, and member of African Biodiversity Network (ABN).

Hon Japhet Mbiuki, MP and former Assistant Minister for Agriculture 2008–2013. Mbiuki is now vice-chairman of the departmental committee of agriculture.

Lilian Muchungi, Community Mobilisation and Advocacy Officer at Green Belt Movement—a national organisation initiated by Professor Wangari Maathai, with 45,000 centres throughout Kenya.

Hon Dr. Victor Munyana, MP, a veterinary surgeon and member of the departmental committee of agriculture.

Dr. Evans Mwangi, Biodiversity expert scientist who co-chaired with Professor Judy Wakhungu, Cabinet Secretary of Environment, water and natural resources, as Kenyan representatives on the ground-breaking, globally important International Assessment of Agricultural Knowledge, Science and Technology for Development (IAASTD) report.

Senator Judith Sijeny, lawyer working mainly on gender issues.

Interviews in Ethiopia

Professor Fetien Abay, Department of Crop and Horticultural Sciences, Mekelle University, Tigray, Ethiopia, and Director of the Institute of Environment, Gender and Development Studies.

Dejene Abeesha, Ministry of Agriculture official.

Dr. Dawit Alemu, Co-ordinator, Agricultural Economics Research-Extension and Farmers Linkage at Ethiopian Institute of Agricultural Research (EIAR).

Kefyaywu Assefa, Agronomist and field contact focal point for malt barley farmers.

Ashenafi Ayenew, Director, Genetic Resource Access and Benefit Sharing directorate at Ethiopian Biodiversity Institute (EBI).

Dr. Yit Barek, Head of seed law development at the Agricultural Transformation Agency.

Dr. Million Belay, Director of Alliance for Food Sovereignty in Africa (AFSA) and Movement for Ecological Learning and Community Action (MELCA).

Tadele Bento, Co-operatives expert and policy development at the Agricultural Transformation Agency.

Wubshet Bernhanu, Country Director of Self Help Africa.

Ghidey Debessu, Advisor to the State Minister of Agriculture.

Seblewongel Deneke, Director, Gender Program at the Ethiopian Agricultural Transformation Agency.

Sue Edwards, Director of Institute for Sustainable Development (ISD).

Regassa Feyissa, Director of Ethio Organic Seed Action and Former Director of Institute of Biodiversity Conservation, now Ethiopian Biodiversity Institute (EBI). A plant physiologist, he is the Ethiopian representative on the Convention on Biological Diversity (CBD).

Dr. Fassil Gebeyhu, General Co-ordinator at African Biodiversity Network (ABN).

Melat Getahun, Program Management Specialist, Economic Growth and Transformation Office (EG&T), United States Agency for International Development (USAID).

Fikremariam Ghion, lawyer at EBI.

Daniel Mekonnen, Ministry of Agriculture (Director of Seeds Section under plant health and regulatory Directorate).

Tigist Mesele, Ministry of Agriculture official.

Wout Soer, Senior Social Protection Specialist at the World Bank.

Mersha Tesfu, Seed policy sector, at the Agricultural Transformation Agency.

Teshome, Ministry of Agriculture (Director of Inputs Directorate).

Tadenya Wakoya, Programme Manager Self Help Africa.

Patricia Wall, Country Representative of Trócaire.

Gary Wallace, Donor Coordinator of Rural Economic Development and Food Security (RED and FS) committee and secretariat member at RED and FS.

Dr. Melaku Worede, Founder of the Institute of Biodiversity Conservation/ Ethiopian Biodiversity Institute (IBC/EBI). Founder of Seeds of Survival, later to become Ethiopian Organic Seed Action (EOSA). Founder of the Ethiopian Academy of Sciences of the African Biodiversity Network (ABN) and former chief scientist at Ministry of Agriculture Ethiopia and United Nations Food and Agriculture Organisation (FAO) Vice Chair.

Geoff Wallis, CEO of Diageo Ethiopia.

Index[1]

[1] Note: Page numbers followed by 'n' refer to notes.

© The Author(s) 2019

C. O'Grady Walshe, *Globalisation and Seed Sovereignty in Sub-Saharan Africa*, International Political Economy Series, https://doi.org/10.1007/978-3-030-12870-8

The manufacturer's authorised representative in the EU is Springer
Nature Customer Service Centre GmbH, Europaplatz 3, 69115 Heidelberg,
Germany. If you have any concerns regarding our products, please
contact ProductSafety@springernature.com

Printed and bound by CPI Group (UK) Ltd, Croydon, CR0 4YY
29/04/2026
02099478-0010